SPECIALTY FOODS

Processing Technology, Quality, and Safety

SPECIALTY FOODS

FOODS

Processing Technology, Quality, and Safety

Edited by Yanyun Zhao

CRC Press
Taylor & Francis Group
Boca Raton London New York

CRC Press is an imprint of the
Taylor & Francis Group, an **informa** business

CRC Press
Taylor & Francis Group
6000 Broken Sound Parkway NW, Suite 300
Boca Raton, FL 33487-2742

First issued in paperback 2019

ISBN-13: 978-1-4398-5423-5 (hbk)
ISBN-13: 978-0-367-38134-9 (pbk)

Library of Congress Cataloging-in-Publication Data

Specialty foods : processing technology, quality, and safety / [edited by] Yanyun Zhao.
    p. cm.
Includes bibliographical references and index.
ISBN 978-1-4398-5423-5 (hardback)
    1. Natural foods industry--United States. 2. Food industry and trade--United States.
3. Natural foods--United States--Quality control. 4. Natural foods industry--United
States--Quality control. I. Zhao, Yanyun, Dr.

TP369.U6S64 2012
338.1'90973--dc23
                            2012009450

Visit the Taylor & Francis Web site at
http://www.taylorandfrancis.com

and the CRC Press Web site at
http://www.crcpress.com

Contents

Preface

Ќ

Specialty food (SF) is a rapidly growing market sector, with an annual growth rate of 8–10%. SF products accounted for 13.1% of all retail food sales in 2010, reaching $55.9 billion in sales. One hundred thirty-eight million American consumers (46%) are purchasing specialty foods.

Although specialty food products are marketed widely, there is now no standard definition of specialty foods. Consumers, even processors and regulators, are confused by the term *specialty foods*. Specialty foods are generally considered unique and high-value food items made in small quantities from high-quality ingredients that offer distinct features to targeted customers who pay a premium price for perceived benefits. Increased production and sale of SF have raised concerns about product quality and safety due to several potential hazards. Many questions with respect to how specialty foods are different from other food sectors, specific processing technologies, controls on quality and microbial safety, etc., remain unanswered. After extensive searching, the only two published books found on the subject of specialty foods focus on business development and marketing of specialty foods. There is no technical book on specialty foods from the food science discipline available.

This book will be the first technical book on specialty foods. It first discusses the unique characteristics of specialty food, the market, and consumer demands and trends, and then focuses on each of the major specialty food segments by covering the key processing technologies, equipment needed, and controls on quality and food safety of the products.

About the Editor

Ḱ

Dr. Yanyun Zhao is a professor in the Department of Food Science and Technology at Oregon State University, with combined research, extension, and education responsibilities in value-added food processing. Her research focuses on developing novel food processing and packaging techniques, and an extension effort delivering education and training programs to the Northwest food processing industry in emerging issues related to fruit and vegetable processing. She teaches fruit and vegetable processing and functional foods courses to undergraduate and graduate students, respectively. Her research in natural antimicrobial edible films/coatings has received worldwide attention and been broadly reported by national and international media. Dr. Zhao has published over 60 peer-reviewed research articles and 17 book chapters, edited a book, *Berry Fruit: Value-Added Product for Health Promotion* (CRC Press, 2007), holds and/or filed 5 patents, and received over 15 USDA competitive grants totaling over $2.5 million. Dr. Zhao has been a member of the Institute of Food Technologists (IFT) since 1991 and has served in various roles, including chair-elect/chair/past-chair of the IFT Fruit and Vegetable Products Division (2008–2011) and IFT Peer Reviewed Communication Committee (2005–2007). She has also served on various university, department, and industry advisory groups, including the OSU Research Council, First International Berry Health Benefit Symposium steering committee chair, department graduate committee chair, department Promotion and Tenure Committee chair, Northwest Food Processors Association Technical Operation Committee, editorial board of the *Journal of Food Processing and Preservation*, and four times a review panelist for the USDA NRI/AFRI research and Integrated Food Safety grant programs.

Contributors

Ḱ

Ramaswamy C. Anantheswaran
Department of Food Science
Pennsylvania State University
University Park, Pennsylvania

B. Douglas Brown
Hershey Food Corporation
Hershey, Pennsylvania

Mark A. Daeschel
Department of Food Science and
  Technology
Oregon State University
Corvallis, Oregon

Jingyun Duan
Food Safety Division
Oregon Department of
  Agriculture
Salem, Oregon

Shawn Fels
Department of Food Science and
  Technology
Oregon State University
Corvallis, Oregon

Lisbeth Meunier Goddik
Department of Food Science and
  Technology
Oregon State University
Corvallis, Oregon

Julie Laughter
Department of Food Science
Pennsylvania State University
University Park, Pennsylvania

Larry Lev
Department of Agricultural and
  Resource Economics
Oregon State University
Corvallis, Oregon

Olga I. Padilla-Zakour
Department of Food Science and
  Technology
Cornell University
Geneva, New York

Hikaru Hanawa Peterson
Department of Agricultural
  Economics
Kansas State University
Manhattan, Kansas

Devin J. Rose
Department of Food Science and
   Technology
University of Nebraska–Lincoln
Lincoln, Nebraska

Melissa Sales
Department of Food Science and
   Technology
Oregon State University
Corvallis, Oregon

Elizabeth K. Sullivan
Department of Food Science and
   Technology
Cornell University
Geneva, New York

Randy W. Worobo
Department of Food Science and
   Technology
Cornell University
Geneva, New York

1

Introduction of Specialty Foods

Yanyun Zhao

Oregon State University
Corvallis, Oregon

Ќ

Contents

Introduction

According to the 2010 Mintel Report of the state of the specialty food industry, the total U.S. market for specialty foods in 2010 reached $70.32 billion, with $55.92 billion of sales at retail, and specialty foods represented 13.1% of all food sales at retail[1] (Figure 1.1). Sales of specialty foods at retail stores have increased by 10.4% since 2008. Americans consume more than $13 billion in specialty foods annually. In 2010, 63% of all consumers reported that they purchase specialty foods, a dramatic leap over the 46% that reported a purchase in 2009, and also higher than the 56% who bought in 2008.[2] Taste (78%), quality (46%), and dietary or health benefits (19%) considerations, coupled with natural ingredients and an absence of preservatives, are major reasons given by consumers for purchasing specialty foods.

Definition of Specialty Foods

The term *specialty foods* has been used widely by food manufacturers, vendors, and distributors and has been used for quite sometime at retail stores. However, how *specialty foods* is defined is vague. There is no universal and standard definition of *specialty foods*, or regulations different from those of conventional foods by the U.S. Food and Drug Administration (FDA) and U.S. Department of Agriculture (USDA). Several organizations have given

FIGURE 1.1
Total U.S. Specialty Food Retail Dollar Sales at Current Prices from 2004 to 2010

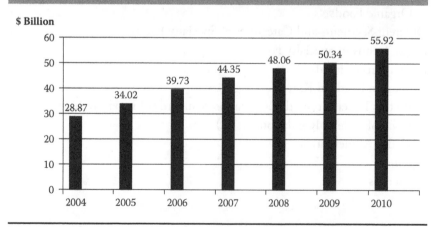

their own definition of *specialty foods*. The best known and most cited one is probably by the National Association for the Specialty Food Trade (NASFT), the leading specialty food association in the United States.

NASFT defines specialty foods as "foods and beverages that exemplify quality and innovation, including artisanal, natural, and local products that are often made by small manufacturers, artisans and entrepreneurs from the U.S. and abroad" (http://www.specialtyfood.com/knowledge-center/industry-info/industry-statistics/).

Another organization that works closely with specialty foods is the National Grocers Association (NGA). According to NGA, "in general, specialty foods can be put into several categories: ethnic foods, regional foods, imported foods and artisan foods. Some foods that may be called specialty in one store may in fact be a staple in another."

Following NASFT's definition, *THE NIBBLE*, a consumer magazine focusing on specialty foods, added its own interpretation of specialty foods (http://www.thenibble.com/nav2/media/kit/FAQ.asp#specialty):

- Specialty foods exist in every category. Most are foods you eat every day: jam, olive oil, vinegar, coffee, tea, meat and poultry, cookies, bread and even butter. Specialty just means better: made with better ingredients and/or artisan techniques. The majority of specialty foods are all natural, made without preservatives.
- Higher quality does not necessarily equate to high cost. Most specialty food items do not cost that much more than their mass-production counterparts.

While there is no universal and standard definition for specialty foods, it is commonly accepted that specialty foods are unique and high-value food items made from high-quality ingredients that offer distinct features to targeted customers who pay a premium price for perceived benefits. Specialty foods may have one or more of the following distinctions:

- Quality of ingredients, manufacturing process, or finished product. Natural, local, organic, eco-friendly, and sustainable are important characteristics for specialty food
- Sensory appeal in flavor, aroma, texture, or appearance
- Presentation from the branding or packaging
- Origin of the products manufactured
- Distribution channel, such as specialty food retail stores or sections within supermarkets and grocery stores

According to a recent survey conducted by a group in the U.S. northwest region to the specialty food processors in Idaho, Oregon, and Washington, uniqueness, special process and design, high quality, and ethnic and exotic origin are the most selected characteristics to describe specialty foods. Of course, the exact characteristics of specialty foods vary depending on the specific category of the specialty food products.

Subsets of Food May Be Included as Specialty Foods and Their Singularity

The use of the term *specialty foods* has been integrated/mingled with ethnic, natural, and gourmet foods. Or, these terms can be included as subsets of specialty foods. There is no clear line to separate these terms of different food categories, although each of them has unique distinctiveness. This section gives specific descriptions about these different food categories with the hope it can provide readers some basic ideas on their uniqueness.

Gourmet Foods

Again, there is no standard, legal definition of *gourmet food*, nor is it differently regulated by the FDA or USDA, but rather used as a marketing term. Gourmet foods are characterized by high quality, accurate preparation, and artistic presentation. *THE NIBBLE* added its definition on gourmet food: "Gourmet food must be defined by complexity and nuance, not by rarity and expense. It is the challenge to the palate and the degree of connoisseurship required to appreciate it that defines 'gourmet,' whether it is Roquefort ice cream, affordable by almost anyone, or golden asetra caviar, accessible to few" (http://www.thenibble.com/nav2/media/kit/FAQ.asp#gourmet).

Vogel (2005) defined gourmet foods as "an expensive, seasonal, nonnative food, perceived as superior, that under the best of circumstances is also of high quality, accurately prepared and presented with artistic flair."[3]

In her book *The New Food Lover's Companion*, Sharon Tyler Herbst described gourmet food as "that which is of the highest quality, perfectly prepared and artfully presented."[4]

Ethnic/Regional Foods

Ethnic or regional foods are a wide variety of foodstuffs, virtually any that can be identified in the public mind with a foreign source or an American minority group. They are prepared or consumed by members of an ethnic group as a manifestation of its ethnicity. Ethnic food in the United States is a $75 billion annual business. Mexican cuisine is and continues to remain the country's most popular ethnic food segment, according to the Datamonitor Report.[5] The next largest food segments are Chinese, kosher, and non-Chinese Asian.[6] Based on the report "The State of the Specialty Food Industry 2011,"[1] Europe remains the primary source for imported specialty foods (80% in 2011), although more products are coming from both Asia and South America. In 2011, Mediterranean and Indian are the most influential emerging cuisines.[1]

Natural Foods

The difference between specialty and natural foods has almost disappeared, as 89% of specialty foods were manufactured or marketed as "nature foods" in 2010.[1] However, it is important to note that the FDA has not formally defined the term *natural*, although it is commonly claimed on the labels of products by manufacturers.

The FDA discussed "natural" claims in the preamble to the 1993 rule establishing nutrient content claim regulations, in which *natural* is defined as meaning "nothing artificial or synthetic (including all color additives regardless of source) has been included in, or has been added to, a food that would not normally be expected to be in the food" (Federal Register, Vol. 58, No. 3, Jan. 6, 1993).

In 1982, the USDA Food Safety and Inspection Service (FSIS) issued Policy Memo 055 on natural claims. It was intended as a guide to manufacturers for developing labeling bearing natural claims that the FSIS would likely find truthful and not misleading. Conditions were specified for using the claim. Based on the USDA FSIS's requirements, two factors have to be considered for natural claims: (1) product does not contain artificial flavor, coloring ingredient, or chemical preservative, or any other artificial or synthetic ingredient, and (2) product and its ingredients are not more than minimally processed. Minimal process means: (1) traditional processes used to make food edible, preserve it, or make it safe (e.g., smoking, roasting, freezing, drying, and fermenting), or (2) physical processes that do not fundamentally alter the raw product or that only separate a whole food into component parts, e.g., ground beef, separating eggs into albumen and yolk, and pressing fruits to produce juices. Relatively severe processes, e.g., acid hydrolysis, are more than minimal. An exception is that case-by-case statements such as "all natural ingredients except for hydrolyzed milk protein" may be approved.

When putting the "natural" claim on the packaging label, *natural* is linked to a brief statement of meaning on the label, e.g., "no more than minimally processed and contains no artificial ingredients." An "all natural ingredients" claim may be used if applicable.

Organic Foods

Although having their own legal definition and regulation, organic foods have also been a subset of specialty food mainly driven by the market. The USDA National Organic Program is a system that is managed in accordance with the Organic Foods Production Act of 1990 and regulations in Title 7, Part 205 of the Code of Federal Regulations (CFR). The National Organic Program develops, implements, and administers national production, handling, and labeling standards. Based on the National Organic Program specification:

- "100% organic" products include all organically produced (raw and processed) ingredients (excluding water and salt). The "100% organic" label may be used, as may the USDA organic seal. The organic certifying agent must be identified on the label, as must the seal.
- USDA certified organics are made with 95% or more organic ingredients. These foods may be labeled as "organic" and carry the USDA organic seal. The name of the certifying agent must appear on the label, although the seal is optional.
- "Made with organic ingredients" means foods may include 70–95% organic ingredients. Up to three of these organic ingredients may be listed on the primary display panel, along with the "made with organic ingredients" tag. The name of the certifying agent must be included; the USDA organic seal can't be used.
- Foods made with less than 70% organic content can include the organic ingredients on the ingredient label. This term can be found on the information panel on applicable products and identifying ingredients. It can't be used on the primary display panel, however, and no seals can be used.

In addition to all natural and organic, local, eco-friendly, and sustainable are other important characteristics of specialty foods.[2] Table 1.1 gives definitions of these terms. According to Tanner (2011),[1] "All-Natural and Local are the buzzwords that interest consumers today. As those become more commonplace, Sustainable will become more prominent."

TABLE 1.1

Definition of Some Other Characteristics Used to Describe Specialty Foods

Term	Definition
Locally sourced	Made with ingredients/materials from less than 200 miles from the geographic location where the foods are sold in retail
Eco-friendly	Uses recyclable packaging or the minimum packaging necessary
Artisanal	Hand-crafted, made in small batches, with high-quality ingredients and techniques
Ethical	Produced with the welfare of animals in mind; products such as cage-free eggs and dolphin-safe tuna are examples
Fair trade	Made with ingredients that take into consideration the health and welfare of workers and farmers. Products often have a fair trade logo
Sustainable	Made with ingredients or packaging that doesn't harm living creatures or the environment

Source. Adapted from Tanner, R., National Association for Specialty Food Trade (NASFT), 2010.

Different Segments and Categories of Specialty Foods

The specialty food market has been divided into more than 40 leading segments, in which condiments; cheese or cheese alternatives; chips, pretzels, and snacks; functional juice, tea, and beverages; coffee, coffee substitutes, and cocoa; and entrées, pizza, and conventional foods have been the leading categories since 2004 (Table 1.2). In 2010,[1]

- Cheese and cheese alternatives were the largest specialty food category, at $3.23 billion, which recorded a 6.1% increase in sales between 2008 and 2010, followed by frozen and refrigerated meats, poultry, and seafood at $1.74 billion, and then chips, pretzels, and snacks at $1.67 billion.
- Functional beverages were the fastest-growing specialty food category, followed by yogurt and kefir, which had 119.5 and 48.9% increases from 2008, respectively.
- The three specialty food categories with the greatest sales penetration are refrigerated sauces, salsas, and dips at 56.9%, teas at 49.2%, and pickles, peppers, olives, and other vegetables at 33.5%.

TABLE 1.2
Sales of Specialty Foods at Retail from 2006 to 2010

Sales of Specialty Foods by Product Categories	2006 $ Million	2007 $ Million	2008 $ Million	2009 $ Million	2010 $ Million	% Change 2006–2010
Cheese and cheese alternatives	2,844	3,142	3,407	3,457	3,231	13.61
Condiments	2,777	2,378	2,429	2,551	1,381	−50.27
Frozen and refrigerated entrées, pizzas, and convenience foods	1,308	1,454	1,704	1,642	1,741	33.10
Chips, pretzels, and snacks	1,074	1,307	1,387	1,509	1,670	55.49
Frozen and refrigerated meats, poultry, and seafood	1,282	1,396	1,543	1,453	1,741	35.80
Frozen desserts	1,271	1,223	1,338	1,303	970	−23.68
Bread and baked goods (frozen and nonfrozen)	899	1,108	1,101	1,260	1,523	69.41
Refrigerated juices and functional beverages	688	877	1,063	1,172	573	−16.72
Coffee, coffee substitutes, and cocoa	928	1,033	1,014	1,066	1,295	39.55
Shelf-stable sauces, salsas, and dips	758	802	870	851	244	−67.81
Cookies and snack bars	677	827	851	838	816	20.53
Yogurt and kefir	450	599	699	830	1,140	153.33
Candy and individual snacks	587	788	811	827	1,003	70.87
Milk, half and half, and cream	633	764	842	822	880	39.02
Shelf-stable pastas	515	576	684	749	459	−10.87
Oils (cooking)	665	718	688	706	692	4.06
Carbonated functional and RTD tea and coffee beverages	594	677	687	700	372	−37.37
Baking mixes, supplies, and flours	530	582	638	687	633	19.43
Shelf-stable fruits and vegetables	566	609	627	672	605	6.89
Teas	557	578	599	627	613	10.05
Cold cereals	457	537	584	571	593	29.76
Shelf-stable juices and functional drinks	541	564	581	566	531	−1.85

TABLE 1.2 (CONTINUED) Sales of Specialty Foods at Retail from 2006 to 2010						
Sales of Specialty Foods by Product Categories	2006 $ Million	2007 $ Million	2008 $ Million	2009 $ Million	2010 $ Million	% Change 2006–2010
Refrigerated sauces, salsas, and dips	403	471	498	549	601	49.13
Crackers and crisp breads	446	477	471	495	515	15.47
Seasonings	370	413	417	473	493	33.24
Soup	375	443	427	452	493	31.47
Beans, grains, and rice	315	345	405	448	379	20.32
Water	403	453	456	428	456	13.15
Nuts, seeds, dried fruits, and trail mixes	332	381	354	370	578	74.10
Energy bars and gels	262	308	339	353	442	68.70
Other dairy	314	336	350	348	244	–22.29
Sweeteners	201	227	245	282	295	46.77
Shelf-stable meats, poultry, and seafood	242	251	240	253	313	29.34
Conserves, jams, and nut butters	188	207	221	238	127	–32.45
Frozen fruits and vegetables	154	175	199	217	256	66.23
Eggs	133	169	174	171	170	27.82
Puddings and shelf-stable desserts	101	118	109	130	135	33.66
Hot cereals	64	71	75	76	77	20.31
Rice cakes	15	17	17	19	20	33.33
Frozen juices and beverages	8	14	8	13	6	–25.00

Source. Modified from Tanner, R., National Association for the Specialty Food Trade, 2010, 2011.

- Overall, specialty foods account for 13.1% of all food sales at retail. The penetration varies significantly by food category, with refrigerated sauces, salsa and dips, and teas boasting more than 40% of category sales.
- Sales of many specialty food categories were booming in 2010 despite the economy.

However, new product introductions fell by 37% between 2008 and 2010 due to the economy's influence (Table 1.3).

TABLE 1.3
New Specialty Food Products Introduced during 2008–2010

Category	2008	2009	2010
Food (subtotal)	**1,283**	**1,193**	**1,104**
Sauces and seasonings	132	107	169
Bakery	171	144	151
Chocolate confectionery	150	155	105
Snacks	180	134	103
Sweet spreads	16	29	33
Dairy	71	26	54
Side dishes	37	53	42
Desserts and ice cream	114	156	78
Processed fish, meat, and egg products	170	156	158
Meals and meal centers	109	84	113
Soup	22	7	13
Savory spreads	17	12	22
Fruit and vegetables	49	70	33
Breakfast cereals	21	15	5
Sweeteners and sugar	5	7	3
Baby food	6	3	6
Beverages (subtotal)	**394**	**329**	**388**
Nonalcoholic beverages	281	249	214
Alcoholic beverages	113	80	174
Total	**1,677**	**1,522**	**1,492**

Source. Modified from Tanner, R., National Association for the Specialty Food Trade, 2010, 2011.

Major Trends in Specialty Foods

Many articles have discussed the trend of specialty foods. The following may be some major trends:[8–10]

- Natural and organic:
  - "Nature" is an overused and sometimes misused claim on the specialty food products.
  - In spite of the fact that a food's naturalness depends upon the definition of *artificial*, it has become a catchphrase for the industry and has given manufacturers a way to distinguish their products from mainstream ones.
  - Appearing on items ranging from breakfast cereal to soda to confections, this term has lent comfort and guidance to many consumers.

- Exotic flavors:
  - Specialty foods are really all about unique and exotic flavors.
  - Wild, exotic fruits, cooking spices, and seasoned salts are more popular.
  - Examples of the product may include less commonly used spices added into condiments served with meat and seafood, spiced fruit jams and jellies, mixture of exotic flavors from salty chocolate to wasabi peanuts, smoked paprika almonds.

- Healthy:
  - Lighter and lower-fat versions of high-end products represent an important trend that is just beginning in specialty foods.
  - Between 2005 and 2009, 2,861 new food and beverage products claimed to promote digestive health, followed by 1,089 products claimed to promote cardiovascular health.
  - Gluten-free products are on the rise as an increasing number of cases of celiac disease are being diagnosed and more people choose to eliminate wheat from their diets as a lifestyle choice.

- Convenience and ease:
  - Because "young gourmets" frequently have limited cooking skills and older gourmets limited time, gourmet frozen/refrigerated entrees, pizzas, and convenience foods are turning in torrid sales.
  - Dessert mixes are another fast-growing specialty food convenience category.
  - Other products just make preparation easier.

- Home cooking:
  - Consumers are cooking at home more and looking for ingredients to "bring restaurant experiences" to their own kitchens.
  - 78% of specialty food consumers like to experiment with new recipes.[1]

- Clean label trend still rising:
  - Fewer additives or ingredients are used in food products.
  - There are decreases in the average number of ingredients in different categories of specialty food products, including dairy products, processed meats, and even pet foods.

- Locally sourced foods:
  - Locally sourced is the most important factor in specialty food purchases. According to NASFT, locally sourced is defined as within 200 miles' distance from the operator.[2]

- New food traditions, brought by new Americans and by increased travel at home and abroad:
  - Watch for growing interest in African foods, Latin American flavors, and homage to American regional cuisines.

Regulations of Foods

The objective of food regulations is to maintain a safe and wholesome food supply. Several state and federal agencies are involved in regulating foods depending on the type of food, how it is prepared or produced, and where it is to be sold. Before a food product is commercialized, it is required that the regulatory agencies should be notified and the product should be inspected. Knowing the regulations that govern food and the facilities used in the production, processing, storage, and distribution of a product is an absolute necessity. It is also the legal responsibility of the manufacturer.

Specialty foods are not regulated differently from other foods. Common food processing standards and practices should be applied in specialty food processing. This section briefly introduces the state and federal agencies that may be involved, their roles, and some important regulations to follow in order to commercialize specialty food products.

State and Federal Regulatory Agencies

The following primary agencies are responsible for regulating food establishments:

- U.S. Food and Drug Administration (FDA)
- U.S. Department of Agriculture (USDA)
- Department of Commence
- National Oceanic and Atmospheric Administration (NOAA)
- Environmental Protection Agency (EPA)
- State Department of Agriculture
- State Public Health Department

It is best to consult knowledgeable sources before investing in a food processing venture as individual states may be subject to different inspections by different state agencies, depending on the type of food manufactured. Table 1.4 lists state and federal agencies regulating different types of foods in general.

TABLE 1.4

General Food Regulations Agencies (Vary Depending on Individual State)

	State Agency	Federal Agency
Milk	Board of Animal Health Department of Agriculture	USDA
Eggs	State egg board	FDA (shell egg) USDA (egg product)
Meat/poultry	Board of Animal Health or Department of Agriculture	USDA
Fruit/vegetable	Department of Health or Department of Agriculture	FDA

Seafood	Department of Agriculture
	Department of Health Services
		FDA
		National Oceanic and Atmospheric
		  Administration (NOAA) of the
		  Department of Commerce
		Environmental Protection Agency
		USDA

Nonalcoholic beverages	Department of Agriculture	FDA

State Agencies The **State Department of Agriculture** is the state agency usually responsible for licensing food establishments and regulating and enforcing food safety as it pertains to food processing, handling, storage, and sale in the state. This usually covers all food prepared for sale to consumers (retail food), distributors, or retailers (wholesale food). The State Department of Agriculture also inspects retail food stores such as groceries, markets, and delis.

The **State Department of Public Health** usually regulates food service establishments such as restaurants, schools, daycares, nursing homes, and hospitals. However, in some states, the State Department of Health is also responsible for licensing and regulating retail and wholesale foods. Check with your state to find out the specific requirements and procedures.

Each state has its own rules and regulations in food establishment, sanitation standards, and distribution of foods. The state retail and food service codes and regulations can be found from the FDA website at http://www.fda.gov/Food/FoodSafety/RetailFoodProtection/FederalStateCooperativePrograms/ucm122814.htm. Note that processed and packaged foods sold across state lines may be regulated and inspected by the FDA.

Federal Agencies The **U.S. Department of Agriculture (USDA)** is a federal agency enforcing laws for meat, poultry, and fish and products that contain these foods as ingredients. If a product contains 3% or more meat products, the processor will be under USDA jurisdiction. Meat and poultry products to be sold in interstate commerce are regulated by the Food Safety and Inspection Service of the USDA (www.fsis.usda.gov). The USDA Agricultural Marketing Service (www.ams.usda.gov) is responsible for developing quality grade standards for agricultural commodities such as milk and shell eggs.

The **Food and Drug Administration (FDA)** (http://www.fda.gov), an agency of the federal government's Department of Health and Human Services, enforces laws requiring all food for human or animal consumption (except meat and poultry) to be safe and appropriately labeled. Foods sold in state do not require FDA approval. However, a food business that has products involved in interstate commerce is subject to compliance with FDA regulations. In essence, a product is involved in interstate commerce when any part of the product (e.g., an ingredient, container, or package) or any part of the product's marketing (e.g., warehousing, distribution, or sale) crosses a state line. The FDA works with the state and local health and agriculture departments to inspect many of the food establishments in state.

The FDA uses the Code of Federal Regulation Title 21 (21 CFR) (discussed below) to regulate the food processing industry. These regulations protect consumers from adulterated and misbranded foods. You can obtain information about 21 CFR from the FDA's district office.

The FDA also requires that owners, operators, or agents in charge of domestic or foreign facilities be registered under the Public Health Security and Bioterrorism Preparedness and Response Act of 2002 (Bioterrorism Act) (www.cfsan.fda.gov/~furls/ovffreg.html). Domestic facilities are required to register whether or not food from the facility enters interstate commerce. Registration provides the FDA with information on the company, the products produced, and the name(s) and contact information for responsible persons in the business. Registration may be accomplished via computer online by going to www.fda.gov and following the links to the forms for completing food registration. A system is also in place to register using printed forms that are available from the local FDA office.

There are many other important state and federal agencies that directly regulate food processing and distribution. Here is a list of some of the federal agencies.

- Agricultural Marketing Service, http://www.ams.usda.gov/
- Agricultural Research Service, one of the world's premier scientific organizations, http://www.ars.usda.gov/main/main.htm

- Bureau of Alcohol, Tobacco and Firearms (ATF), http://www.atf.treas. gov/alcohol/index.htm
- Center for Science in the Public Interest foodborne illness outbreak database, http://www.cspinet.org/foodsafety/outbreak/pathogen.php
- Cooperative Development, http://www.rurdev.usda.gov/rbs/coops/ cswhat.htm
- Food and Drug Administration (FDA), http://www.fda.gov
- FDA food facility registration information, www.cfsan.fda.gov/~furls/ ovffreg.html
- Food and Nutrition Service, http://www.fns.usda.gov/fns/
- Food assistance program, provides easy access to the best food and nutrition information from across the federal government, http://www.nutrition.gov
- USDA Food Safety and Inspection Service (FSIS), http://www.fsis.usda.gov
- USDA Rural Development, http://www.rurdev.usda.gov/index.html
- Internal Revenue Service (IRS), enforcement of tax laws, tax collection, http://www.irs.gov/smallbiz/index.htm
- Seafood Inspection Service, part of the U.S. Department of Commerce/ National Oceanic and Atmospheric Administration, http://seafood.nmfs.gov
- U.S. Small Business Administration (SBA), offers a variety of programs through which small businesses can secure loans from bank; the loans are guaranteed by the SBA but administered by the lending institution, http://www.sba.gov

Factors Determining the Regulation Standards of Food Products[11]

Several key elements determine how a food product will be regulated. As stated in the above section, the type of a food, e.g., fresh produce, milk, seafood, meat, eggs, and the marketing or distribution channel applied (retail or wholesale) determine the specific regulation standards required. In addition, a food product is regulated depending on its acid strength, which is measured in terms of pH. The pH measurement of acid strength is reported on a scale of 0 to 14.0, with neutral being 7.0. When a food product has a pH above 4.6, it is called low-acid food, including foods such as meat, poultry, seafood, milk, and fresh vegetables (except for tomatoes). Acid foods have a pH of 4.6 or lower. These include foods like jams and jellies, sauces, most salad dressings, and most fruits. Also, some foods are inherently low in acid, such as cucumbers, okra, cauliflower, and peppers, but acid is added as part of the process, to make the final pH of the product below 4.6, i.e., a pickled product. These foods are classified as acidified foods and are regulated under a different set of guidelines. Acidified and low-acid foods can be produced only under an approved process authority, which is a professional with specific qualifications, and at a separate inspected location or by a food manufacturing facility.

FDA Draft Guidance for Industry: Acidified Foods was issued in October 2010, and can be found at the FDA website at http://www.fda. gov/downloads/Food/GuidanceComplianceRegulatoryInformation/ GuidanceDocuments/AcidifiedandLow-AcidCannedFoods/UCM227099. pdf. Detailed discussions about the processing and regulations of acidified foods are discussed in Chapter 5.

Code of Federal Regulations (CFR)

The Code of Federal Regulations (CFR) is the codification of the general and permanent rules published in the Federal Register by the executive departments and agencies of the federal government. It is divided into 50 titles that represent broad areas subject to federal regulation. Each volume of the CFR is updated once each calendar year and is issued on a quarterly basis.

Each title is divided into chapters, which usually bear the name of the issuing agency. Each chapter is further subdivided into parts that cover specific regulatory areas. Large parts may be subdivided into subparts. All parts are organized in sections, and most citations in the CFR are provided at the section level.

Parts 1 to 1499 of CFR Title 21, Food and Drugs, are administered by the U.S. Food and Drug Administration and the U.S. Drug Enforcement Administration. Title 21 regulates the food processing industry. Here are examples of specific regulations described in the different parts of CFR Title 21.

- Part 1—General provisions
- Part 101—General rules
- Parts 108, 113, 114—Low-acid (pH 4.6 and above) canned food and acidified food
- Part 108.25—Acidified food
- Part 110—Current Good Manufacturing Practices (GMP) in manufacturing, processing, packing, or holding human food
- Part 120—Hazard Analysis and Critical Control Point (HACCP) systems
- Part 120.6—Sanitation Standard Operating Procedures (SSOP)

- Parts 130–169—Standardized foods (e.g., standard of identity, standard of quality, fill of containers, imitation food)
- Part 184—Direct food substances affirmed as generally recognized as safe (GRAS)
- Part 601—Licensing
- Part 801—Labeling

For accessing CFR Title 21 via the Internet, the FDA website maintains a searchable site for 21 CFR for retrieving any 21 CFR chapter and part (http://www.accessdata.fda.gov/scripts/cdrh/cfdocs/cfcfr/cfrsearch.cfm). It can also be accessed through the National Archives and Record Administration website (http://www.access.gpo.gov/cgi-bin/cfrassemble.cgi?title=200121).

Codex Alimentarius

The Codex Alimentarius (Latin for "food book") Commission (http://www.codexalimentarius.net/) was created in 1963 by the Food and Agriculture Organization (FAO) of the United Nation and World Health Organization (WHO) to develop international food standards, guidelines, and related texts such as codes of practice under the Joint FAO/WHO Food Standards Program. The main purposes of this program are protecting the health of consumers, ensuring fair trade practices in the food trade, and promoting coordination of all food standards work undertaken by international governmental and nongovernmental organizations.

The general texts include the following:

- Food labeling (general standard, guidelines on nutrition labeling, guidelines on labeling claims)
- Food additives (general standard, including authorized uses, specifications for food-grade chemicals)
- Contaminants in foods (general standard, tolerances for specific contaminants, including radionuclides, aflatoxins, and other mycotoxins)
- Pesticide and veterinary chemical residues in foods (maximum residue limits)
- Risk assessment procedures for determining the safety of foods derived from biotechnology (DNA-modified plants, DNA-modified microorganisms, allergens)
- Food hygiene (general principles, codes of hygienic practice in specific industries or food handling establishments, guidelines for the use of the Hazard Analysis and Critical Control Point (HACCP) system)
- Methods of analysis and sampling

Codex Alimentarius is a collection of internationally recognized standards, codes of practice, guidelines, and other recommendations relating to foods, food production, and food safety. It contains over 300 standards (http://www.codexalimentarius.net/web/standard_list.jsp) for specific products. Here are some specific standards:

- Meat products (fresh, frozen, processed meats and poultry)
- Fish and fishery products (marine, freshwater, and aquaculture)
- Milk and milk products
- Foods for special dietary uses (including infant formula and baby foods)
- Fresh and processed vegetables, fruits, and fruit juices
- Cereals and derived products, dried legumes
- Fats, oils, and derived products such as margarine
- Miscellaneous food products (chocolate, sugar, honey, mineral water)

Summary

The specialty food industry continuously grows, mainly driven by consumers' demands on high quality, exotic flavor and taste, elegant appearance, and health and environment consciousness. Gourmet, natural, and ethnic foods are the major subsets of specialty foods. Other important characteristics of specialty foods may include locally grown, artisanal, and unique processing. Quality, flavor, and taste are everything about specialty foods. By now, there is neither standard definition nor different regulation on specialty foods. They are regulated following the same standards applied on conventional food products. It is the responsibility of processors, vendors, and distributors to follow the state and federal regulatory requirements for ensuring quality and food safety of specialty food products.

References

1. Tanner, R. 2011. The state of the specialty food industry 2011. National Association for Specialty Food Trade (NASFT).
2. Tanner, R. 2010. Today's specialty food consumer 2010. National Association for Specialty Food Trade (NASFT).
3. Vogel, M.R. 2005. Gourmet food. Food for thought. January 12, 2005. http://www.foodreference.com/html/art-gourmet-food.html (accessed October 1, 2011).
4. Herbst, S.T. 2001. *The new food lover's companion*. 3rd ed. Barron's Educational Series.

5. Datamonitor. 2005. A Datamonitor report: Insights into tomorrow's ethnic food and drinks consumers. Product code: DMCM2363.
6. Silverstein, B. 2009. Ethnic food brands: A guide to the world on a shelf. http://www.brandchannel.com/features_effect.asp?pf_id=477 (accessed October 1, 2011).
7. Tanner, R. 2010. The state of the specialty food industry 2010. National Association for Specialty Food Trade (NASFT).
8. Rotkovitz, M. 2010. The top 5 specialty food trends for 2010. Gourmet foods, flavors, and specialty finds that'll be big in 2010. About.com. http://gourmetfood.about.com/od/wheretobuygourmetfoods/a/foodtrends2010.htm (accessed October 1, 2011).
9. Sloan, A.E. 2008. Prime time for fancy foods. *Food Technology* 62(7): 26–36.
10. Tanner, R. 2010. Specialty food trends: 2010 and beyond. National Association for Specialty Food Trade (NASFT). http://www.ctfarmrisk.uconn.edu/documents/NASFTTrendsPresentation.pdf (accessed October 1, 2011).
11. Morris, W.C. 2005. Getting started in a food manufacturing business in Tennessee. UT Extension publication PB1399. E12-4815-00-004-05.
12. Shoukas, D. 2010. Food trends. *Specialty Food Magazine* 107.

2

Specialty Food Markets and Marketing

Larry Lev

Oregon State University
Corvallis, Oregon

Hikaru Hanawa Peterson

Kansas State University
Manhattan, Kansas

Ḱ

Contents

Introduction

This chapter examines the diverse specialty foods marketplace and the strategies and practices that entrepreneurs need to understand to be successful specialty food marketers. First is an overview of the specialty food market. The next section considers the most important consumer segments within the specialty food marketplace and their preferences. Next, the main actors in specialty food supply chains and what they do are explained. The final section provides structured ways for food manufacturers to use this background information to begin the process of formulating a marketing plan.

The Specialty Food Market: The Big Picture

Many consumers in affluent countries take it for granted that their calorie needs will be met on a daily basis. Consequently, they focus their attention on food quality rather than food quantity. For these consumers, food not only sustains them, but also provides pleasure and allows them to express their personal values. Even in difficult economic periods, many of these consumers regard food as an affordable luxury.

Food, similar to other consumption goods, can be characterized by three different types of product attributes.[1,2] Search attributes are product characteristics that consumers use to identify and choose products at the point of purchase, including price, brand, and packaging. Experience attributes are what consumers discover about the product after their purchase through consumption. Taste is the most prominent experience attribute, but certain aspects of convenience are also discovered through experience. Credence attributes are product characteristics that cannot be easily verified by consumers and must be revealed through labeling or some other means of communicating with consumers. Examples include whether ingredients are natural or organic, who produced the product, and where it originated. Clearly, all attributes are important as consumers evaluate products, but an increasing focus has been placed on credence attributes to differentiate among specialty food products.

As defined in Chapter 1, specialty foods are superior to mainstream products and can be described using adjectives such as premium, fancy, or gourmet, in one or more of these attributes. Sometimes, specialty foods are referred to as value-added, meaning that the product is differentiated from a generic food product by one or more of the production, packaging, processing, or marketing processes.

In almost all cases, specialty food products cost more. There are numerous reasons and justifications for these price premiums. One overall way of thinking about this is to focus on the extra amount consumers are willing to pay for these specific distinguishing attributes. For example, how much more will consumers pay for an artisanal, local cheese as compared to generic cheese produced and shipped in from afar? Specialty food manufacturers also need to consider whether the higher consumer prices will cover the additional costs necessary to provide the distinguishing attributes. For example, it might be quite costly to produce organic blueberry jam, because people have to be hired to weed instead of applying chemical pesticide. Price premiums can also be achieved by products with limited availability.

The overall food market is typically categorized by food groups, that is, fruits and vegetables, meats, dairy and eggs, snacks, condiments, prepared foods, and so on. Specialty foods can encompass any of these food categories, including unprocessed farm products that are distinguished by sustainability claims or packaged in nongeneric ways. The specialty food market is segmented in multiple ways. One way is to categorize them by common characteristics. Thus, you can think of ethnic/imported foods, natural/organic foods, and so on. Alternatively, the market can be defined by product format, such as ready-to-eat meals, canned, frozen, shelf stable, refrigerated, and fresh deli/prepared.

In 2010, specialty food sales in the United States were $70.3 billion (not including the sales at Wal-Mart in the retail figure).[3] The 80% of specialty food products sold by retailers ($55.9 billion) accounted for 13.1% of all food sales at retail in 2010. There have been about 1,100 to 1,200 new specialty food products with stock-keeping units (SKUs) introduced annually during the last few years. Many specialty foods have penetrated the broader mainstream market. Once exotic, refried beans and soy sauce are found among traditional food items on most grocery shelves. Many restaurant goers are familiar with dessert items such as tiramisu and crème brûlée.

As reported by the 2010 Mintel/NASFT survey, 40% of total specialty food sales at retail are price lookup (PLU) items such as baked goods, prepared foods, and meat.[3,4] For the remaining 60%, which consists of sales

TABLE 2.1
Sales of Specialty Foods at Retail in 2008 and 2010 ($ Million)

Product Category	2008	2010	% of Change, 2008–2010
Cheese and Cheese Alternatives	3046	3231	6.1
Frozen and Refrigerated Meats, Poultry, and Seafood	1577	1741	10.3
Chips, Pretzels, and Snacks	1442	1670	15.8
Bread and Baked Goods	1311	1523	16.2
Condiments, Dressings, and Marinades	1259	1381	9.7
Coffee, Coffee Substitutes, and Cocoa	1037	1295	24.9
Yogurt and Kefir	706	1140	61.5
Frozen Lunch and Dinner Entrées	1058	1073	1.4
Candy and Individual Snacks	883	1003	13.6
Frozen Desserts	909	970	6.7
Milk	854	880	3.1
Cookies and Snack Bars	814	816	0.3
Oils and Vinegars	680	692	1.8
Pickles, Peppers, Olives, and Other Vegetables	629	669	6.5
Baking Mixes, Supplies, and Flour	569	633	11.3
Teas	569	613	7.8
Shelf Stable Fruits and Vegetables	545	605	11.0
Refrigerated Salsas and Dips	492	601	22.1
Cold Cereals	588	593	0.9
Nuts, Seeds, Dried Fruit, and Trail Mixes	441	578	31.1

Source. Tanner, R., National Association for Specialty Food Trade (NASFT), 2011.

of SKU items, Table 2.1 lists the top 20 specialty food product categories in 2010 according to SPINS, a market research and consulting firm for the natural products industry. It is a highly fragmented market with the top eight categories accounting for only 37% of the SKU sales share.

In relative terms, European and Japanese specialty food markets surpass the United States, as food products in those markets have historically been highly differentiated. Origin labels serve as an important means of branding products and assuring quality. Many have become globally known, including Roquefort and Stilton cheeses, prosciutto hams from different Italian regions, and many wines, including champagne and burgundy. News from Japan that a single melon sold for nearly $10,000 seems quite outlandish, but it simply reflects a cultural tradition to respect and mark the beginning of its limited seasonal availability.

Specialty Food Product Consumers: Who They Are and What They Want

Key demographic trends that are expected to shape the U.S. food market in the next decade include an aging and more diverse population, rising income, higher educational attainment, improved diet and health knowledge, and the growing popularity of eating out.[5] The desire to seek out specialty foods is fueled and directed at least in part by food programs on TV. According to a Hartman Group poll, 61% watch cooking shows on TV, with 21% doing so frequently.[6] As a component of affordable luxury, more consumers are entertaining at home in response to reduced occasions for eating away from home.

In a 2010 national survey of 1,500 adults by Mintel and NASFT, 63% participants were specialty food buyers, which is comparable to the prerecession level in 2006.[7] These specialty food consumers visit grocery outlets more frequently than the average grocery shopper. Specialty food consumers spend about $90 per week on food they prepare and eat at home, with about 25% of that spent on specialty foods. Not surprisingly, they spend more time eating meals and report a strong interest in enjoying the eating experience. More than 80% of specialty food consumers indicated that they like to experiment with new recipes.

The specialty food consumers seek out high-end products and ingredients; in short, they look for the quality attributes that define products as specialty foods. Taste is cited as the number one reason why consumers buy specialty foods, according to the Mintel/NASFT survey.[3] Specialty food consumers also select products consistent with their personal values.

While a myriad of attributes can define specialty foods, some product attributes are most commonly seen in successfully marketed specialty foods in the United States. The top two are convenience and healthiness. Specialty foods by definition are rich, indulgent, and many require labor- and time-intensive preparation. But, today's specialty food consumers prefer products with seemingly contradictory combinations of attributes, such as ready-made meals or products requiring minimum preparation, but also fresh, or indulgent, but also low-fat or low-carb. In particular, specialty food consumers are more likely than the average consumer to consider health reasons in their buying decisions and seek out healthful or "better for you" options.[7] Other key attributes include natural/organic (primarily implying additive-free) and local/sustainable. It should be noted that the only reason why food safety is not mentioned as a primary desirable attribute by specialty food consumers is because many assume that providing safe food is an absolute requirement. Any health-related concern or recall will severely challenge specialty food businesses of all sizes.

The popularity of specific characteristics changes frequently and some trends are short-lived. In 2010, the most commonly purchased international/ethnic foods were Italian, Hispanic, and Asian foods, while Mediterranean, Latin (besides Mexican), and Indian were singled out as emerging cuisines by importers.[4,7] The number of new products marketed as gluten-free nearly doubled in 2010 from 2009.[3] The best ways of keeping up on current trends are scanning food magazines, watching TV food programs, and reading web blogs.

Consumer Segments

The globalization of food preferences is seen across the urban-rural spectrum and demographics, suggesting that the rise in consumer interest toward specialty foods is widespread.[6] Still, certain consumer segments emerge as most interested in specialty foods. As an example of how specific some of these target populations can be, Mintel identified the most active specialty food consumers as being age 18 through 34, in a household earning over $75,000 annually, Hispanic, and living in the northeastern part of the country. Overall, consumers in the 25 to 34 age group and those in households earning over $100,000 annually spend the most dollars on specialty foods.[7]

Among generations, generation Y is considered an avid specialty food consumer segment. The number of these individuals born in the middle of the 1980s and after (currently ages 15–30) is estimated to be 70 to 76 million. They are viewed as tech savvy, family and friend focused, and attention craving.[8] Their willingness to experiment, appreciation of food as an art form, limited cooking skills, and time-pressed lifestyles mean they are a prime audience for new specialty food products.[9]

Another large subgrouping of specialty food consumers is called lifestyle of health and sustainability (LOHAS) consumers. Also known as ethical consumers or consumer activists, these consumers take ethical and environmental concerns into consideration when making their purchase decisions. The worldwide estimated size of this market segment includes about 20% of the U.S. population (over 41 million), at least 33% of the European population, and about 30% of the population in Japan.[10]

Ethnic groups in the U.S. population constitute segments of loyal consumers for multicultural foods. The current largest ethnic groups are 48.4 million Latinos, 41.8 million African Americans, and 15.5 million Asians. By 2050, the Latino and African American consumers are projected to reach 133 million and 65 million, respectively.[11] Particularly for the Latinos, the projected growth of their purchasing power is notable. Their buying power, which was estimated to be $978 billion in 2009, is expected to rise to $1.3 trillion in 2014.[12]

Geographical Distribution

Specialty food consumers are concentrated in a limited number of locations. In general, urban areas and coastal regions tend to attract the younger adults and ethnic populations who are the most attractive specialty food market segments. Table 2.2 lists the 40 "primary trade areas" for specialty foods organized by state, with overall rankings indicated in parentheses.[13]

Other clues for geographic distribution of specialty food consumers can be found through activity measures for certain specialty food segments. For example, Local Harvest, Inc. maps farmers' markets and other venues that offer access to local foods (http://www.localharvest.org/) and Slow Food USA, a national organization that supports food lovers and food activists, lists all local chapters (http://www.slowfoodusa.org/). Similarly, the Buy Fresh Buy Local campaign maintains a list of current chapters (http://www.foodroutes.org/bfbl-chapters.jsp#chapter-list). These are not perfect indicators but certainly will help identify where specialty food consumers are most likely to be found.

Specialty Food Product Supply Chain: An Overview

Specialty food products that are produced from manufacturers (including farmers) can reach consumers through various supply chains. Most specialty food products in the United States travel long distances and pass through multiple hands as they travel from manufacturers to consumers. Others are sold directly to consumers in local markets. For new manufacturers to begin analyzing their opportunities, it is critical to understand the state of the supply chain and the range of options.

Manufacturers and Importers

Currently, 70% of specialty food products sold in the United States are manufactured in the United States, and the remaining 30% are imported. Most products are sold to consumers through retailers, brokers, and distributors,

**TABLE 2.2
Top 40 Specialty Food Markets
Organized by State (Rankings in
Parentheses)**

California	Los Angeles (2)
	San Francisco (4)
	San Diego (16)
	Riverside/San Bernardino (25)
	Sacramento (31)
Colorado	Denver (17)
Connecticut	Hartford (29)
District of Columbia	Washington, DC (6)
Florida	Miami/Ft. Lauderdale (10)
	Tampa/St. Petersburg (28)
	West Palm Beach (40)
Georgia	Atlanta (19)
Hawaii	Honolulu (27)
Illinois	Chicago (3)
Indiana	Indianapolis (33)
Louisiana	New Orleans (30)
Massachusetts	Boston (9)
Maryland	Baltimore (18)
Michigan	Detroit (5)
Minnesota	Minneapolis/St. Paul (15)
Missouri	St. Louis (21)
	Kansas City (24)
North Carolina	Charlotte (37)
New Mexico	Phoenix (13)
New York	New York (1)
	Buffalo (39)
Ohio	Cleveland/Akron (14)
	Cincinnati (26)
	Columbus (32)
Oklahoma	Oklahoma City (34)
Oregon	Portland (23)
Pennsylvania	Philadelphia (7)
	Pittsburgh (20)
	Allentown/Bethlehem (38)
Texas	Houston (8)
	Dallas/Ft. Worth (11)
	San Antonio (35)
Utah	Salt Lake City (36)
Washington	Seattle/Tacoma (12)
Wisconsin	Milwaukee (22)

while a small proportion of products are sold directly to consumers. The following overview of the specialty food product supply chain is based on the 2010 Mintel/NASFT survey,[4] where 113 manufacturers, 39 importers, 16 distributors, and 35 brokers of specialty food participated. It is not reported what percentage of the industry is represented by these respondents.

The scale of suppliers can be measured by the number of SKUs and annual sales. Manufacturers of specialty food are diverse. In 2009, a quarter of the manufacturers responding to the survey produced 10 or fewer SKUs, while 21% manufactured over 100 SKUs. About 40% grossed $500,000 or less in annual sales, while 20% sold more than $5 million. The largest share of sales was generated through selling to retail, either directly (42%) or through distributors (28%). The majority of remaining sales were generated equally through direct sales to consumers (13%) and to food service establishments (12%), either directly (6%) or through distributors (6%). Food service establishments that bought specialty food products from manufacturers included hotels, fine and casual dining, institutions (such as schools and hospitals), and conference centers.

Similarly, both small- and large-scale importers bring in specialty food products from overseas. A quarter of the importers responding to the survey carried 50 or fewer SKUs, while 15% carried more than 500 SKUs. A third grossed over $5 million, while 13% sold under $500,000. Regarding sourcing regions, 85% sourced from Europe, followed by 33% from Asia and 21% from South America. It was the most typical for an importer to source from two to four countries. Similar to manufacturers, the largest share of sales was generated from sales to retailers, and only 2% of sales were generated from selling directly to consumers.

The NAFST website (http://www.specialtyfood.com/SOI2011definitions/) provides a useful listing of production definitions and examples of firms that supply products in all of the specialty food categories. Examples illustrate that many well-known and relatively large firms participate in the specialty food marketplace. Furthermore, many of the best-known specialty food manufacturers started as independent firms and were subsequently purchased by larger mainstream firms, such as Kashi (owned by Kellogg's), Cascadian Farms (General Foods), and Odwalla (Coke) (see other examples in this diagram: https://www.msu.edu/~howardp./organicindustry.html).

Given this competition, any manufacturer wishing to compete must recognize "there are three ways to be attractive to the consumer: be cheaper, be better, or be unique. Being inexpensive will make marketing easier, but most niche marketers are more likely to have high quality or unique products."[14] Alternative marketing strategies are further discussed in the final section.

Retailers, Food Service, and Direct Sales

Tanner[3] provides the information needed to understand the sales outlets for specialty food products. Eighty percent are sold to retail food outlets, and the other 20% are sold to the food service sector. Among retail food outlets, mainstream supermarkets dominate, with 72% of the sales, followed by sales through specialty food stores (19.3%) and natural food stores (8.4%). Retailers typically mark up the products they sell by 35 to 50%. Outlets such as farmers markets, farm stands, special fairs, and Internet sales (either direct or through third-party sites) are not broken out in these statistics but may be the most appropriate starting places for many new manufacturers.

For food manufacturers, three advantages of direct sales are (1) receiving all or most of the consumer dollars, (2) close contact with the consumers, which allows for collecting valuable reactions and insights from them, and (3) no minimum volume hurdles to have their products handled by retailers or distributors. While direct sales can be a great starting place, they handle small quantities of product and remain outside the mainstream for food distribution. Many manufacturers will recognize that to achieve greater sales they will eventually need other ways of selling to consumers. The brief case study on page 34 provides an example of using direct sales as a springboard to a larger business.

Brokers and Distributors

While some sales are made directly to a final consumer and others go straight from the manufacturer to a retail outlet, manufacturers will find it worthwhile in many instances to hire the services of brokers or distributors. Even smaller manufacturers that lack the experience or marketing resources to effectively navigate the complex retailer and food service distributions systems may need to consider hiring these intermediaries. This decision requires a careful calculation of the trade-offs between the opportunities to access additional markets and the generally reduced per unit returns.

Brokers are the people that food manufacturers hire to sell products to retail clients. Brokers typically focus on a specific segment of the retail trade, such as supermarkets, natural food stores, or specialty stores, and they typically handle products for many different manufacturers. Two key things to know about brokers are that (1) they do not take ownership of the product and (2) they work for a commission on wholesale revenues—typically 5 to 15%. Manufacturers often find that using the services of a broker may represent the most cost-effective way to expand into regional or national markets.[15,16] About 80% of brokers who responded to the Mintel/NAFST survey[4] handled between 500 and 5,000 SKUs, while 54% averaged $1 million to $5 million in annual sales. More than half serviced over 250 stores, while about a quarter serviced fewer than 50 stores.

In contrast to brokers, food distributors purchase products from a manufacturer and then sell them to an entity further along the food distribution channel. Manufacturers choose to make use of distributors to gain their ability to access new clients and devote the time necessary to service new accounts. Distributors earn from 10 to 35% of the wholesale price for their efforts. Food distributors provide the following services:[17,18]

- Purchase food products
- Inventory products
- Take sales orders
- Deliver product to retailers
- Stock and rotate products on retailer shelves
- Coordinate in-store demonstrations
- Distribute point-of-purchase materials to retailers
- Educate store staff about products
- Invoice customers and collect payments

By hiring a distributor, small food manufacturers can compete with larger manufacturers who have this expertise in-house. Distributors seek new products that complement the line of products that they already handle, thereby making their own business more attractive. Identifying and beginning to work with a distributor can be challenging, because distributors prefer to add well-established brands rather than unknown products to their line-up. A superior product will eventually find a willing distributor.

Many of the distributors who responded to the Mintel/NAFST survey were large, with 38% handling more than 1,000 SKUs, generating annual sales over $10 million, and servicing over 1,000 stores.[4] At the other end of the spectrum, 19% carried fewer than 50 SKUs, 13% grossed under $200,000, and 19% serviced fewer than 50 stores. The largest share of sales was generated through specialty food stores (36%), followed by supermarkets (20%), food service/restaurants (14%), and natural food stores (10%).

Developing a Marketing Strategy

This section looks more closely at what an individual food manufacturer must do to carve out a viable niche within the specialty food distribution system.

One overriding message is that food manufacturers, who have focused primarily on product development, must shift toward a marketing mindset. As explained below, carefully thinking through the firm's "marketing mix," or what is often called the 4Ps, provides one means for moving toward a viable marketing plan for the enterprise. Other critical strategic elements related to product development and production processes are treated in later chapters of this book.

Market Research

Most beginning food manufacturers do not have the resources or know-how to conduct the sophisticated market research efforts of the larger firms. Nonetheless, all food manufacturers can collect information that will help them to better understand their potential customers and markets. Key market analysis elements are[19,20]

- Estimates of the current and potential size of the market for the product
- Whether the market is divided by types of consumers or types of quality of products
- Seasonality, if any, of the market
- Analysis of the competition

Much of this information can be collected by reading widely, observing carefully, and interviewing thoughtfully. Sometimes it will be necessary to supplement this qualitative information with more rigorous quantitative market research, which the firm can either do itself or obtain by hiring an expert. By the end of this data gathering process, the entrepreneur should have a better understanding of the firm's target customers, how it intends to reach them, and the competition that will be faced in the marketplace.

The Marketing Mix

The marketing mix or 4Ps represent a very simplified version of a marketing plan (examples of marketing plans are available at http://www.mplans.com/). The focus from the start rests on potential consumers of the firm's products and what they want.

Product is the first P. If manufacturers have been focusing on their own product in isolation from the marketplace, this P should serve as a reminder that to compete, products must be both better than the competition's and something that consumers actually want. Food manufacturers who choose to sell through intermediaries need to recognize that their products will have to fit the needs of distributors and retailers as well. The entire set of attributes needs to be fleshed out as products are first designed and then produced.

Place, or where the products will be sold, is the second P. As discussed above, specialty food products are sold in diverse types of outlets, and each food manufacturer needs to target the best set of sales locations. Hall[13] gives the example of a Cajun-style food product that will do better as a gift/souvenir item than as a supermarket product. The size and sophistication of the business and the specific nature of products will determine which outlets to target. What works for the new entrepreneur still eager to receive maximum customer feedback is quite different from what is possible for larger manufacturers. While selling to a national supermarket chain is probably impossible for a new firm, smaller, independent retailers may provide a better fit. Over time, as the number of places grows or the geographic area expands, the specialty food entrepreneur may decide to transition to working with brokers or distributors to service these new outlets.

Getting the *price* right is critical for the success of all specialty food businesses. The simpler case is for manufacturers who direct market their products. As a start, these manufacturers must have a firm grasp of *all* the costs required to produce and distribute their products. But just knowing production and distribution costs is in many instances not sufficient. That is, effective pricing is not based on a strict markup from cost. The specialty food producer must also know the prices of competitive products and how consumers view the firm's product and competitive products.

When products pass through the hands of intermediaries before reaching consumers, additional cost components must be built into the calculations. In general, a manufacturer must work backwards from a competitive consumer price to calculate the firm's selling price. The manufacturer and all of its distribution channel partners must be able to earn adequate returns for the product to succeed. Although every product is different, a product's price increases with every exchange of hands. The manufacturer must know what commission the broker requires and how much the distributor and retailer will mark up the product.

Beamon and Johnson[21] provide a detailed example of how the retail dollar is split up. Assume a manufacturer sells a product to a distributor, using a broker. The distributor sells to retailers, with a final retail price of $0.99 to the consumer. The manufacturer's total cost to produce the product is $0.49. With a 30% markup, the manufacturer sells the product to the distributor for $0.64. Subtracting the 5% ($0.03) commission to the broker, the manufacturer's gross revenue is $0.61 per unit sold. The distributor purchases the product for $0.64 and marks it up 15% to $0.74. Retailers then purchase the product for $0.74, mark it up 35%, and sell it to the consumer for $0.99.

The retail dollar selling price is broken down as follows (these numbers are based on average markups):

Manufacturer's total cost	$0.49
Manufacturer's 30% markup	$0.15
Broker's 5% commission	$0.03 (paid by manufacturer and not added to price)
Distributor's 15% markup	$0.10
Retailer's 35% markup	$0.25
Retail price	$0.99

Hall[13] (pp. 71–77) and Beamon and Johnson[21] (pp. 10–12) provide additional examples and worksheets that help illustrate the terminology and calculations needed to understand how prices are determined in the distribution system.

Promotion is the fourth and final P. Promotional goals may include stimulating demand, differentiating the firm's product from the competition's, or simply building awareness. The promotional efforts that individual manufacturers choose run the full gamut from face-to-face methods appropriate in a local farmers market, to participation in major food shows, and expanded use of the Internet. Using the Internet provides a set of intriguing possibilities for the small specialty food products firm:[22]

- Allows small companies to compete with other companies both locally and nationally
- Offers a convenient way of doing business transactions, with no restrictions on hours of operation
- Offers an inexpensive way for small firms to compete with larger companies by having their products available worldwide

Starting Small and Growing

One example of starting small and constructing a viable specialty food manufacturer is Ruby Jewel, a successful premium ice cream treat producer in Portland, Oregon (http://www.rubyjewel.net/index.php). Lisa Herlinger introduces the story of her firm by observing, "We all have a calling, and mine just happened to be ice cream." Her initial inspiration began with the observation of a successful ice cream innovation in Los Angeles—putting quality ice cream between two cookies. Later, after relocating to Portland, she was inspired by working in a farmers' market to think about developing a similar high-quality product using local ingredients to target market customers. Her initial farmers' market success led to a much larger business than she ever imagined, with sales in natural food stores and specialty outlets in 11 states and a freestanding ice cream shop. At each step in the process, Lisa carefully researched her options and worked with the right experts before moving forward.

Summary

The expansion of the specialty food marketplace has been driven by consumers who view these superior products as affordable luxuries. Even in difficult economic times, many specialty products flourish as consumers choose to spend their entertainment dollars on cooking at home. In the marketplace, specialty food consumers seek out products with seemingly contradictory combinations of attributes, such as ready-made meals or products requiring minimum preparation but that are also fresh, or indulgent products that are also low-fat or low-carb. Specialty food consumers share with the general population a desire for convenient food products but differ from average consumers in their preference for healthful or "better for you" options.

Specialty food manufacturers face numerous options in selling their products. Small or start-up manufacturers may choose to market their products directly to consumers and retail clients. As they grow, however, most specialty food manufacturers use the services of distributors and brokers in order to sell to more mainstream food supply chains.

The marketing mix or 4Ps approach provides food manufacturers with the basic elements needed to establish a marketing strategy for their product. They must define the product, decide on the places it will be sold, calculate the price they will charge, and establish a promotion plan that is appropriate. While understanding the marketplace and following this structured approach to marketing does not guarantee success, it will improve the odds.

References

1. Nelson, P. 1970. Information and consumer behavior. *Journal of Political Economy* 78: 311–29.
2. Darby, M.R., and E. Karni. 1973. Free competition and the optimal amount of fraud. *Journal of Law and Economics* 16: 67–88.
3. Tanner, R. 2011. The state of the specialty food industry 2011. NAFST. *Specialty Food Magazine.*
4. Mintel. 2010. Specialty foods—The NASFT state of the industry report—The Market—U.S. Available at Kansas State University library in Mintel report database.

5. Lin, B.-H., J.N. Variyam, J. Allshouse, and J. Cromartie. 2003. *Food and agricultural commodity consumption in the United States: Looking ahead to 2020.* Agricultural Economic Report 820. Food and Rural Economics Division, Economic Research Service, U.S. Department of Agriculture.

6. Hartman Group. 2007. *Multicultural foods: Armchair cooking, travel and the new ethnic dining frontier.*

7. Mintel. 2010. Specialty foods—The NASFT state of the industry report: The consumer—U.S. Available at Kansas State University library in Mintel report database.

8. Kane, S. 2011. Generation Y. About.com. http://legalcareers.about.com/od/practicetips/a/GenerationY.htm.

9. Sloan, A.E. 2009. State of the industry report: Prime time for fancy foods. In *Sell your specialty food.* New York: Kaplan Publishing, Appendix O, pp. 273–86.

10. Harding, A. 2010. The rise and growth of LOHAS 'lifestyles of health and sustainability.' Presentation at Green Unplugged, June 7. http://www.slideshare.net/GreenUnplugged/andrew-harding (accessed March 2011).

11. U.S. Census Bureau. 2009. U.S. population projections. http://www.census.gov/population/www/projections/index.html.

12. Geisler, M. 2011. Ethnic foods market profile. Agricultural Marketing Resource Center, Iowa State University.http://www.agmrc.org/markets_industries/food/ethnic_foods_market_profile.cfm.

13. Hall, S. 2009. *Sell your specialty food.* New York: Kaplan Publishing.

14. Thilmany, D., and J. Grannis. 1998. Marketing food products: Direct sales vs. distributors and brokers. Agricultural Marketing Report 98-04. Fort Collins: Colorado State University Extension Service. http://dare.colostate.edu/pubs/amr98-04.pdf.

15. Beamon, J.A., and A.J. Johnson. 2006. *Using food brokers in the Northwest.* EM 8922. Corvallis: Oregon State University Extension Service. http://ir.library.oregonstate.edu/xmlui/bitstream/handle/1957/20488/em8922.pdf.

16. Appalachian Center for Economic Networks (ACEnet). 2006. Food ventures: Choosing and using brokers. http://www.acenetworks.org/warehouse/Choosing_and_Using_Brokers.pdf.

17. Beamon, J.A., and A.J. Johnson. 2006. *Using food distributors in the Northwest.* EM 8923. Corvallis: Oregon State University Extension Service. http://ir.library.oregonstate.edu/xmlui/bitstream/handle/1957/20487/em8923.pdf.

18. Appalachian Center for Economic Networks (ACEnet). 2006. Selling to retailers. Available at http://www.acenetworks.org/warehouse/Selling_to_Retailers.pdf.

19. Holz-Clause, M. 2009. Conducting market research. Agricultural Marketing Resource Center. http://www.agmrc.org/business_development/starting_a_business/marketbusiness_assessment/articles/conducting_market_research.cfm.

20. Appalachian Center for Economic Networks (ACEnet). 2006. Researching your product category. http://www.acenetworks.org/warehouse/Researching_Your_Product_Category.pdf.

21. Beamon, J.A., and A.J. Johnson. 2006. *Food distribution channel overview.* EM 8921. Corvallis: Oregon State University Extension Service. http://ir.library.oregonstate.edu/xmlui/bitstream/handle/1957/20443/em8921.pdf.

22. Holz-Clause, M. 2009. Marketing on the Internet. Agricultural Marketing Resource Center. http://www.agmrc.org/business_development/operating_a_business/direct_marketing/articles/marketing_on_the_internet.cfm.

3

Food Safety of Specialty Foods and Their Controls

Jingyun Duan

Oregon Department of Agriculture
Salem, Oregon

Ḱ

Contents

Introduction

While taste and quality are important to specialty foods, more important is making sure that the consumer always gets a safe and wholesome food product. To protect the consumer and ensure the products are the safest possible, the specialty food processors must take all the necessary steps to prevent or eliminate potential food safety hazards at all stages of their operation.

Food safety is defined as "the assurance that food will not cause harm to the consumer when it is prepared and/or eaten according to its intended use" by the Joint Food and Agriculture Organization of the United Nations (FAO) and World Health Organization (WHO) Food Standards Program.[1] Food safety hazard refers to any physical, chemical, or biological contaminations that may cause a food to be unsafe for human consumption. Table 3.1 summaries the physical, chemical, and biological hazards related with food safety.

To ensure food safety, controls should be applied from the beginning to the end at all stages of food processing to reduce any potential hazards as a means of prevention rather than intervention of finished products. To prevent the physical hazard, processors must identify the sources and types of materials that can be physical hazards in foods, and determine the types of controls to minimize the potential for physical hazards. Successful chemical control programs include training employees to follow safe handling and application procedures for sanitation; making standard practices for staff to properly clean and remove all chemical residues from food contact surfaces; storing chemicals in designated areas away from food, food ingredients, packaging, and food contact surfaces; receiving incoming materials and raw ingredients from reputable suppliers that effectively control chemical hazards; ensuring restricted ingredients and additives are correctly measured; and following good storage practice. The basic food safety principles for controlling biological hazard include utilizing effective processing steps to reduce the numbers of pathogenic microorganisms in food to a safe level, practicing good personal hygiene, controlling time and temperature to limit the amount of time that potentially hazardous foods are held in the temperature danger zone during preparation, and preventing cross-contamination from food to food, from people to food, and from environment to food.

TABLE 3.1
Physical, Chemical, and Biological Hazards Related with Food Safety

	Definition	Source	Examples
Physical hazard	Any extraneous object or foreign matter in a food item that may cause illness or injury to the consumers	Raw materials, badly maintained facilities and equipment, improper production procedures, and poor employee practices	Wood, plastic, metal, glass, bones, whole spices and herbs, insects and rodents, their parts and excreta, personal items
Chemical hazard	Toxic substances and any other compounds that may lead to acute food-borne illness, chemical poisoning, or food allergy	Allergens, metals, chemicals, and poisonous substances	**"Big 8" food allergens:** Milk, egg, fish, crustacean shellfish, peanut, tree nut, wheat, soybean **Heavy metals:** Mercury, copper, lead, zinc, tin **Chemicals:** Cleaning chemicals, sanitizers, lubricants, pesticides, fungicides, insecticides, fertilizers, intentional food additives (added in excess), unintentional additives **Poisonous substances:** Botulin.um toxin, patulin, paralytic shellfish poisoning (PSP) toxins
Biological hazards	Include parasites, viruses, and bacteria, which may lead to food spoilage or food-borne illness	Animals (manure, animal living spaces, and carcasses), people (food handlers, pickers, packers, and consumers), environment (contaminated water, air, and plants)	**Parasites:** *Cyclospora cayetanensis, Cryptosporidium parvum, Giardia lamblia* **Viruses:** Hepatitis A virus, norovirus, rotavirus **Bacteria:** *Salmonella* spp., *Escherichia coli* O157:H7, *Listeria monocytogenes, Shigella* spp.

Source.   Adapted from Duan, J., Y. Zhao, and M. Daeschel, *Ensuring Food Safety in Specialty Foods Production*, OSU Extension publication (EM 9036), 2011.

This chapter introduces the major food safety programs that have been well recognized internationally and provides examples of implementing food safety controls in some specialty food products.

Potential Food Safety Hazards in the Production of Specialty Food Products

Increased specialty food production has raised food safety concerns in this unique food category. In addition to the common food safety hazards associated with mainstream food processing, other potential hazards related with specialty foods may include

- Many specialty foods are claimed to be "all natural" with "no added preservatives." While this can be of great appeal to consumers, barriers preventing contamination by microorganisms, especially food-borne pathogens, may be lacking.
- Most specialty foods are manufactured in small-scale operations and some processors are newly started family businesses, which may have limited training and experience in safe food production practices and may be unaware of licensing and inspection requirements. Thus, limited training and education in food safety may be an issue.
- Some uncommon and poorly characterized ingredients and processing procedures may be used in the production of certain specialty foods, which may lead to potential allergen or microbial safety problems.

Hundreds of food recalls are reported each year in the United States, causing huge economic loss and the loss of consumer confidence. Many food recalls have been associated with specialty food products. For example, over 10 specialty food recalls were reported during the period of January to March 2011, which included a pumpkin chipotle roasting sauce recall due to undeclared wheat flour, a wheat-free and gluten-free French bread pizza recall due to potential contamination with *Listeria monocytogenes*, and a coconut water recall due to potential growth of mold.[3]

Major Food Safety Programs

The potential hazards and reported specialty food recalls necessitate the implementation of effective food safety programs for ensuring safe production of specialty foods. The programs that are used most often to ensure food safety and quality of processed foods are Good Manufacturing Practices (GMPs), Sanitation Standard Operation Procedures (SSOPs), and Hazard Analysis and Critical Control Points (HACCP). In addition, other food safety programs that have been well recognized internationally may include Global Food Safety Initiative (GFSI), British Retail Consortium (BRC), ISO 22000, and Safety Quality Food (SQF). The following sections give brief descriptions about these programs.

Global Food Safety Initiative (GFSI)

GFSI was launched in May 2000. It is managed by the consumer goods forum, the only independent global network for consumer goods retailers and manufacturers worldwide. As a non-profit-making foundation, the GFSI aims at continuous improvement in food safety management systems to ensure confidence in the delivery of safe food to consumers.[4,5]

The GFSI is working through the implementation and maintenance of the GFSI Guidance Document to recognize and compare food safety standards worldwide in order to encourage mutual recognition of these standards and to avoid multiplication of effort.[6] The GFSI also communicates to stakeholders about system equivalence, provides a forum for debate with international standards organizations and interested parties, and helps and encourages retailers and other stakeholders to share knowledge and strategy for food safety through different projects. The GFSI does not undertake any accreditation or certification activities, but encourages third-party audits against the standards it recognizes in order to encourage efficiency and to allow suppliers to direct their resources to continuous improvement in food safety and quality.

The GFSI Guidance Document contains commonly agreed criteria for food safety standards, against which any standard can be measured, compared, and ultimately recognized by the GFSI. It is not a standard itself, but a framework in which food safety management schemes can be benchmarked. The Guidance Document, 5th edition,[7] represents food safety management best practice in the form of key elements for food production as requirements for food safety management schemes, a conforming food safety management standard, and the delivery of food safety management systems. It also provides guidance on how to seek compliance for existing systems owners and on the operation of certification processes.

The GFSI operates a 10-step benchmark procedure to ascertain whether a standard and its certification system can demonstrate conformity against the

Guidance Document.[8] In addition, the GFSI ensures that the benchmark procedure is implemented in an independent, impartial, technically competent, and transparent manner. Under the umbrella of the GFSI, eight major retailers, including Carrefour, Tesco, ICA, Metro, Migros, Ahold, Wal-Mart, and Delhaize, have come to a common acceptance of GFSI-benchmarked food safety schemes. As more retailers accept a GFSI-benchmarked standard as proof of compliance to a food safety system, manufacturers could reduce the duplication of audits and improve cost efficiency in the supply chain.

While GFSI encourages food businesses to choose GFSI-recognized schemes, specialty food manufacturers can make individual choices whether or not to implement the program. Although choosing a GFSI-recognized scheme may be a large investment for a business initially, the number of audits to be reduced after implementation would significantly reduce the financial burden of multiple audits. In addition, when an outside customer requires specific audits, under the GFSI framework only one scheme would be required.

Good Agricultural Practices (GAPs) and GlobalGAPs

Broadly defined, GAPs is an approach that addresses environmental, economic, and social sustainability for on-farm processes and results in safe and healthy food and nonfood agricultural products.[9] GAPs applications have been developed not only by the food industry and producer organizations, but also government and nongovernmental organizations, aiming to fulfill trade and government regulatory requirements, and to meet farmers' and transformers' needs and specific requirements. The objectives of GAPs are to

- Ensure safety and quality of produce in the food chain
- Capture new market advantages by modifying supply chain governance
- Improve natural resources use, workers' health, and working conditions
- Create new market opportunities for farmers and exporters in developing countries

When focusing on food safety and quality, GAPs guidelines are directed toward four primary components of production and processing: soil, water, hands, and surfaces, as described in Table 3.2.[10]

GlobalGAPs, originally named EurepGAPs, is a key reference for GAPs in the global marketplace, by setting voluntary standards for the certification of production processes of agricultural products.[11] It aims at establishing one standard for GAPs with different product applications capable of fitting to the whole of global agriculture. The GlobalGAPs standard is a pre-farm-gate

TABLE 3.2
GAPs Guidelines on Four Primary Components
of Production and Processing

Components	Guidelines
Soil	• Maintain clean soil: Reduce the risk of introducing microbial contaminants from the soil into the produce during stages of growth and harvesting. • Apply proper manure management with attention to composting, storage, and proper application timing. • Minimize the presence of animals in production fields and packing areas.
Water	• For washing, cooling, and processing: Must be of drinkable quality, i.e., potable. • For irrigation and foliar applications: Must be tested to ensure the compliance with minimum quality levels. • Regular testing of water sources must be performed to verify the water quality, and the testing records must be available. • Buffer areas and fencing are needed to protect ground and surface water from runoff and animal contamination.
Hands	• Have clean hands: Manage the human impact on product quality and safety. • Clean hands applies to workers and the use of good personal hygiene during production and processing. • Provide clean and appropriately stocked restroom and hand washing facilities to field. • Apply proper hand washing techniques by processing employees.
Surfaces	• Maintain clean surfaces: Ensure that all produce contact surfaces during harvest and processing, including harvest equipment and containers, transport bins, knives and other utensils, sorting and packaging tables, product packaging, and storage areas, are properly cleaned and sanitized on a regular basis. • Apply effective cleaning and sanitation procedures to achieve the correct level of hygiene in food handling or production facilities: debris removal, use of detergent solutions, rinsing with water, disinfection where necessary, and dry cleaning.

Source. Adapted from Iowa State University Extension, On-Farm Food Safety: Guide to Good Agricultural Practices (GAPS). Retrieved from http://www.extension.iastate.edu/Publications/PM1974a.pdf.

standard designed to reassure consumers about how food is produced on the farm by minimizing detrimental environmental impacts of farming operations, reducing the use of chemical inputs, and ensuring a responsible approach to worker health and safety as well as animal welfare. The certificate covers the process of the certified product from farm inputs like feed or seedlings and all the farming activities until the product leaves the farm. The scope of GlobalGAP currently covers the production of fruit, vegetables, combinable crops, green coffee, tea, flowers and ornamentals, cattle and sheep, dairy, calf/young beef, pigs, poultry, turkey, and aquaculture.

The U.S. Department of Agriculture (USDA) operates an audit/certification program to verify that farms use GAPs or Good Handling Practices (GHPs).[12] The program is intended to assess a participant's efforts to minimize the risk of contamination of fresh fruits, vegetables, nuts, and miscellaneous commodities by microbial pathogens based upon a U.S. Food and Drug Administration (FDA) publication entitled *Guide to Minimize Microbial Food Safety Hazards for Fresh Fruits and Vegetables*. The USDA guidelines focus on food safety, and do not address topics such as animal welfare, biodiversity, or the use of antibiotics and hormones. Sections of the audit checklist that are covered in the USDA's audit/certification program include the following:

General questions:
Part 1—Farm review
Part 2—Field harvest and field packing activities
Part 3—House packing facility
Part 4—Storage and transportation
Part 5—(Not in use at this time)
Part 6—Wholesale distribution center/terminal warehouse
Part 7—Preventative food defense procedures

The USDA GAP/GHP audit program is voluntary. The specialty food processors could request the audit, and when they meet the requirements of the audit program, they will receive a certificate (valid for 1 year) and are listed on the USDA Agricultural Marketing Service website (http://www.ams.usda.gov/AMSv1.0/GAPGHPAuditVerificationProgram).

Hazard Analysis and Critical Control Point (HACCP)

The HACCP program is a systematic preventive approach to minimize food safety risks through the analysis and control of biological, chemical, and physical hazards from raw material production, procurement, and handling, to manufacturing, distribution, and consumption of the finished product. The concept of HACCP was developed in the 1960s by the Pillsbury Company while working with the U.S. National Aeronautics and Space Administration (NASA) and the U.S. Army Laboratories as a means of ensuring the safety of food produced for the U.S. space program.[13] Since then, HACCP has become gradually recognized as a valuable approach that could provide the greatest assurance of safety while reducing dependence on finished product testing. It has been widely used by industry since the late 1970s, and is now universally recognized as the best system for ensuring food safety. The FDA and the USDA have issued regulations that made HACCP mandatory for seafood, juice, and meat products as an effective approach to food safety and protecting public health.[14] The HACCP system has also been implemented in many other countries, such as

Europe, Canada, Australia, and New Zealand, and is a high-priority program under Codex Alimentarius, the world food standards authority.

The format of HACCP plans varies with the different products and processes. It is important that the unique conditions within each facility be considered when developing the HACCP plan. Five preliminary tasks need to be accomplished prior to the application of the HACCP principles to a specific product and process, and are described in Table 3.3.

After these five preliminary tasks have been completed, the seven principles of HACCP are applied to develop the HACPP plan.

Principle 1: Conduct a hazard analysis. The HACCP team should conduct a thorough hazard analysis to list all of the hazards that are of such significance that they are reasonably likely to cause injury or illness if not effectively controlled. Two stages are involved in conducting a hazard analysis: hazard identification and hazard evaluation. During the first stage, a list of potential biological, chemical, or physical hazards that may be introduced

TABLE 3.3
Preliminary Tasks that Need to Be Accomplished Prior to Applying HACCP Principles

Tasks	Description
1. Assemble the HACCP team	Consist of individuals who have specific knowledge and expertise appropriate to the product and process. Local personnel who are involved in the operation should also be included in the team, as they are more familiar with the variability and limitations of the operation. Assistance from outside experts who are knowledgeable in the potential hazards associated with the product and the process may be needed as well.
2. Describe the food and its distribution	Include ingredients, processing methods, packaging, storage conditions, and method of distribution.
3. Describe the intended use and consumers of the food	May be the general public or a particular segment of the population, e.g., infants, elderly, immunocompromised individuals, military, hospital patients.
4. Develop a flow diagram which describes the process	Provide a clear, simple outline of all the steps involved in the operation. Steps preceding and following the specified operation can also be included.
5. Verify the flow diagram	On-site review of the operation should be performed to verify the accuracy and completeness of the flow diagram and make modifications where appropriate.

Source. Adapted from the U.S. Food and Drug Administration, Hazard Analysis and Critical Control Point Principles and Application Guidelines, 1997. Retrieved from http://www.fda.gov/Food/FoodSafety/HazardAnalysisCriticalControlPointsHACCP/HACCPPrinciplesApplicationGuidelines/default.htm.

or increased at each step in the production process is developed by reviewing the ingredients used in the product, the activities conducted at each step in the process and the equipment used, the final product and its method of storage and distribution, and the intended use and consumers of the product. In the hazard evaluation stage, potential hazards must be addressed based on the severity of the potential hazard and its likely occurrence. The nature of food, its method of preparation, transportation, storage, and the intended consumers' susceptibility to a potential hazard should be considered to determine how each of these factors may influence the likely occurrence and severity of the hazard being controlled.

Upon completion of the hazard analysis, the HACCP team must list the hazards associated with each step in the production along with any measures that are used to control the hazards. More than one control measure may be required for a specific hazard and more than one hazard may be controlled by a specific control measure.

Principle 2: Determine the critical control points (CCPs). A critical control point (CCP) is defined as a step in a food process where control can be applied and is essential to prevent or eliminate a food safety hazard or reduce it to an acceptable level. Examples of CCPs may include thermal processing, chilling, testing ingredients for chemical residues, pH control, and testing products for microbial contaminants. Complete and accurate identification of CCPs is essential to controlling food safety hazards. The determination of a CCP can be facilitated by the use of a decision tree, which indicates a logic reasoning approach. However, the decision tree is merely a tool, and it should be used as guidance, but not a substitute for expert knowledge when determining CCPs. Examples of decision trees are given in Figures 3.1 and 3.2.

Principle 3: Establish critical limits. A critical limit, used to distinguish between safe and unsafe operating conditions at a CCP, is a maximum or minimum value to which a biological, chemical, or physical parameter must be controlled at a CCP to prevent, eliminate, or reduce to an acceptable level the occurrence of a food safety hazard. In some cases more than one critical limit will be elaborated at a particular CCP. Criteria often used include temperature, time, moisture level, pH, water activity (a_w), titratable acidity, salt concentration, available chlorine, and sensory parameters such as visual appearance and texture. All critical limits must be scientifically based and derived from sources such as regulatory standards and guidelines, literature surveys, experimental results, and experts.

Principle 4: Establish monitoring procedures. Monitoring is the scheduled observations or measurements to assess whether a CCP is under

FIGURE 3.1
Example I of a CCP Decision Tree[14]

Q 1. Does this step involve a hazard of sufficient likelihood of
occurrence and severity to warrant its control?

    ↓   ↓

   YES   NO →Not a CCP

    ↓

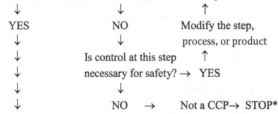

Q 2. Does a control measure for the hazard exist at this step?

    ↓   ↓   ↑

  YES   NO   Modify the step,

   ↓   ↓   process, or product

   ↓   Is control at this step   ↑

   ↓   necessary for safety? → YES

   ↓   ↓

   ↓   NO   →   Not a CCP→ STOP*

Q 3. Is control at this step necessary to prevent, eliminate, or reduce the
risk of the hazard to consumers?

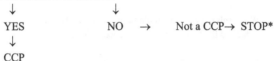

    ↓   ↓

  YES   NO   →   Not a CCP→ STOP*

   ↓

  CCP

Source. From the U.S. Food and Drug Administration, Hazard Analysis and Critical Control Point Principles and Application Guidelines, 1997. Retrieved from http://www.fda.gov/Food/FoodSafety/HazardAnalysisCriticalControlPointsHACCP/HACCPPrinciplesApplicationGuidelines/default.htm.

control and to produce an accurate record for future use in verification. Monitoring facilitates tracking of the operation, determines if there is loss of control and a deviation occurs at a CCP, and provides written documentation for use in verification. Ideally, continuous monitoring should be provided to ensure control of the process. If continuous monitoring is not feasible, it is necessary to ensure the amount or frequency of monitoring is sufficient to guarantee the CCP is in control. Most monitoring procedures need to be done rapidly because they relate to on-line, real-time processes. Therefore, physical and chemical measurements, such as visual observations and measurement of temperature, time, pH, and moisture level, are often preferred because they may be done rapidly and can often indicate the microbiological control of the product.

Principle 5: Establish corrective actions. Specific corrective actions must be developed for each CCP in the HACCP system to ensure appropriate actions will be taken when monitoring indicates a deviation from an established critical limit. Corrective actions must include

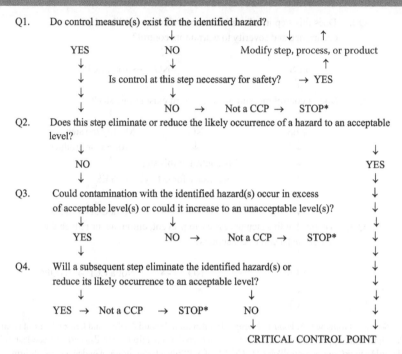

FIGURE 3.2
Example II of a CCP Decision Tree[14]

Q1.   Do control measure(s) exist for the identified hazard?

    YES   NO   Modify step, process, or product

         Is control at this step necessary for safety?   → YES

         NO   →   Not a CCP →   STOP*

Q2.   Does this step eliminate or reduce the likely occurrence of a hazard to an acceptable level?

    NO   YES

Q3.   Could contamination with the identified hazard(s) occur in excess of acceptable level(s) or could it increase to an unacceptable level(s)?

    YES   NO   →   Not a CCP →   STOP*

Q4.   Will a subsequent step eliminate the identified hazard(s) or reduce its likely occurrence to an acceptable level?

    YES →   Not a CCP   →   STOP*   NO

                                 CRITICAL CONTROL POINT

Source. From the U.S. Food and Drug Administration, Hazard Analysis and Critical Control Point Principles and Application Guidelines, 1997. Retrieved from http://www.fda.gov/Food/FoodSafety/ HazardAnalysisCriticalControlPointsHACCP/HACCPPrinciplesApplicationGuidelines/default.htm.

determination and correction of the cause of noncompliance, proper disposition of the affected product, and record of the corrective actions that have been taken. This part of the HACCP program should specify what is done when a deviation occurs, who is responsible for implementing the corrective actions, and that a record will be developed and maintained of the actions.

Principle 6: Establish verification procedures. Verification consists of those activities, other than monitoring, that determine the validity of the HACCP plan and show that the system is operating according to the plan. One important aspect of verification is the initial validation of the HACCP plan, which determines that the plan is scientifically and technically sound, all hazards have been identified, and these hazards will be effectively controlled if the HACCP plan is properly implemented. Another aspect of verification is to verify that the validated plan is being followed correctly by activities such as calibration of process monitoring instruments, direct observation of monitoring activities and corrective actions, random sampling and analysis, and review of records. Verification activities can be carried out by individuals within a company, third-party experts, and regulatory agencies. The frequency of verification should be sufficient to confirm that the HACCP system is working effectively to ensure the control of the hazards.

Principle 7: Establish record-keeping and documentation procedures. Efficient and accurate record keeping and documentation are essential to the application of a HACCP plan. It is important that all plants maintain certain documents, including its hazard analysis and written HACCP plan, and records documenting the monitoring of critical control points, critical limits, verification activities, and the handling of processing deviations.

The FDA and the USDA have issued regulations that make HACCP mandatory for seafood, juice, and meat products as an effective approach to food safety and protecting public health. The use of HACCP is limited within small companies. However, working through all the stages of a HACCP system would bring a thorough understanding of food safety issues affecting specialty food business and confidence in the products, which allows specialty food processors to challenge the legitimacy of demands from enforcement officers, external auditors, and others, while at the same time tapping their knowledge and experience to help review and refine the system.[15]

British Retail Consortium (BRC)

The British Retail Consortium (BRC), which was formed in January 1992 when the British Retailers' Association and the Retail Consortium merged, is one of the lead trade associations in the United Kingdom.[16] In 1998, the BRC developed and introduced the BRC Food Technical Standard to assist retailers and brand owners in producing food products of consistent safety and quality and assisting with their "due diligence" defense.[17] In a short period of time, this standard was widely adopted not just throughout the UK but around the world, and has evolved into a global standard and been regarded as the benchmark for best practice in the food industry. Subsequently, the BRC published the first issue of the Packaging Standard in 2002, followed by Consumer Products Standard in August 2003, and finally by the BRC Global Standard—Storage and Distribution in August 2006. In 2009, the BRC developed the Global Standard For Consumer Products North America edition in association with the Retail Industry Leaders Association (RILA) in the United States.

The BRC Global Standards are a suite of four industry-leading technical standards that specify requirements for organizations to enable the production, packaging, storage, and distribution of safe food and consumer products.

1. Global Standard for Food Safety. As the first standard to be approved by the GFSI as part of its process for mutual recognition of food safety standards, BRC Global Standard for food safety continues to be the most widely used of the GFSI standards, with over 10,000 certificated sites worldwide. Fundamental requirements by this standard include senior management participation, the adoption of HACCP, a documented quality management system, and control of factory environmental standards, processes, and personnel.[18]
2. Global Standard for Packaging and Packaging Materials. The standard provides greater focus on quality and functional aspects of packaging and complements the established requirements of factory hygiene. It is now a leading global standard adopted worldwide by major retailers, manufacturers, and packaging businesses.
3. Global Standard for Consumer Products. The standard defines the requirements for the production and supply of safe and legal consumer products (retailer branded, branded, or unbranded) of consistent quality.
4. Global Standard for Storage and Distribution. The standard is designed to complete the chain between food manufacturer and the retailer and ensure best practice in handling storage and distribution of products.

The BRC Standards have now become the internationally recognized mark of excellence with over 14,000 certificated suppliers in over 100 countries through a network of over 90 accredited and BRC-recognized certification bodies. For food producers, meeting the standards allows them to demonstrate their compliance with industry requirements, and also allows them to be listed on a secure database of certified suppliers. Certification to the BRC Standards reduces the number of supplier audits, which greatly benefits the retailers, food producers, importers, caterers, ingredient suppliers, and food service industry.

ISO 22000

ISO 22000, *Food Safety Management System—Requirements on the Organization of Food-Chain*, was issued in September 2005 by the International Organization for Standardization, and is the first management system standard on food safety to go beyond the recommendations put forward in 1993 by the Codex Alimentarius Commission. It not only endorses the Codex Alimentarius recommendations, but also attempts to fill the gaps and inconsistencies brought to light by 13 years of accumulated experience with HACCP. The key innovations of ISO 22000 include the following:[19]

- Responsibility and authority of food safety team leader: Organizing the team's training and work, ensuring the implementation and updating of the system, reporting to management and communicating.
- Interactive communication: External communication relating to food safety hazards throughout the food chain (upstream and downstream); internal communication to ensure that the HACCP team is informed in real time of all changes likely to affect the system.
- Human resources: Demonstrated competence of the HACCP team members and the staff having an impact on food safety is required.
- Prerequisite programs (PRPs): The company should select and implement appropriate good hygiene practices by itself (instead of merely applying those imposed upon it).
- Hazard identification and determination of acceptable levels: Taking into account the various stages in the food chain where hazards can occur; determining acceptable levels in the finished product.
- Selection and assessment of control measures: Selection of control measures associated with hazards assessed as requiring control; assessment of the effectiveness of control measures; method for assigning these control measures either to the HACCP plan (conventional CCP) or to operational PRPs (new concept).
- Establishment of the operational prerequisite programs (PRPs): Establishing a monitoring system for the control measures assigned to the operational PRPs.
- Validation of control measure combinations: Prior validation of the effectiveness of the control measure combinations to ensure observance of the predefined acceptable level for the relevant hazard.
- Evaluation of individual verification results: Systematic review of individual results of the plan and verification.
- Analysis of results of verification activities: Analyzing and overall reviewing the implementation, operation, and efficiency of the system and the trends in terms of hazard control, with management reporting.

These innovations incorporate appropriate prerequisite programs and HACCP principles, thus ensuring effective prerequisite programs are in place for assuring a clean sanitary environment; an HACCP plan is developed for identifying, preventing, and eliminating food safety hazards; and processes are established and implemented for managing food safety throughout the organization.

ISO 22000 is an auditable standard with clear requirements. It integrates and harmonizes various existing national and industry-based certification schemes, and contributes to a better understanding and further development of HACCP. ISO 22000 can be applied on its own or in combination with other management system standards to any organizations directly or indirectly involved in the food chain. Food safety management systems that conform to ISO 22000 can be certified, which provides an effective means to communicate with stakeholders and other interested parties, and demonstrates food safety commitment under corporate governance, corporate responsibility, and financial reporting requirements.

Safe Quality Food (SQF)

The Safety Quality Food (SQF) program is a leading, global food safety and quality certification and management system. It was launched by SQF Institute in 1994 and is the only certification system recognized by GFSI. The program provides independent certification to meet the needs of suppliers and buyers worldwide and ensure their compliance with food safety regulations in both domestic and global markets at all stages of the supply chain.[20]

The SQF program comprises two standards based on the type of food supplier:[21]

1. SQF 1000 Code. Based on the principles of HACCP and designed for producers of primary food products, including those involved in horticulture, produce, meat, poultry, dairy, eggs, coffee and cereal production, fishing and aquaculture. It enables them to meet product trace, regulatory, food safety, and commercial quality criteria, and also allows primary producers to demonstrate that they can supply food that is safe and meets the quality specified by a customer.
2. SQF 2000 Code. A complete HACCP system and designed for manufacturers, distributors, and brokers of food and beverage products. It covers the identification of food safety and quality risks, and the validation and monitoring of control measures, and can be used by all sectors of the food industry.

Both SQF codes are divided into three certification levels, which provide suppliers with an opportunity for continuous improvement and allow every supplier, from the smallest farmer to the largest manufacturer, to be eligible for SQF certification.[19] Table 3.4 describes the three certification levels of the SQF program.

TABLE 3.4 Certification Levels of SQF Program	
Certification Levels	*Description*
Level 1 Food safety fundamentals	Suppliers must establish prerequisite programs that incorporate fundamental food safety controls essential to providing a sound foundation for the production and manufacturing of safe food.
Level 2 Certified HACCP-based food safety plans	Suppliers must complete and document a food safety risk assessment of the product and process by using the HACCP method, and an action plan to eliminate, prevent, or reduce food safety hazards.
Level 3 Comprehensive food safety and quality management system	Suppliers must complete and document a food quality assessment of the product and its associated process, to identify the controls needed to ensure a consistent level of quality.

Source. Adapted from the SQF Institute, One World, One Standard. Retrieved from http://www. sqfi.com/about-sqfi.

A level 1 certificate is appropriate for low-risk products, while level 2 is the minimum level for suppliers of high-risk products, and both of them are prerequisites to gaining a level 3 SQF certificate. At all levels, suppliers are required to meet buyer product specifications and the regulatory requirements of the countries where they operate or export products. Suppliers are issued with an SQF certificate indicating the level of certification achieved, and after achieving level 3, an SQF-certified supplier is authorized to use the SQF 1000 or 2000 certification trademark to indicate its status as an SQF quality-certified supplier.

Based on their business size and the product nature, specialty food processors could chose different levels of SQF certification to enhance and demonstrate the safety of their products.

Sanitation Standard Operation Procedures (SSOPs)

Sanitation Standard Operating Procedures (SSOPs) are mandatory for all meat and poultry processing plants (9 CFR Part 416) and for all food processing plants subject to HACCP (21 CFR Part 120.6). Although specific protocols may vary from facility to facility, SSOPs provide specific, step-by-step written procedures necessary to ensure sanitary handling of foods. They include written steps for cleaning and sanitizing, before, during, and after operations, to prevent direct contamination or adulteration of product(s).[21]

Like the HACCP plan, SSOPs are specific to each plant, but may be similar to plants in the same or a similar industry.[22] These documents describe procedures for eight sanitation conditions,[23] as summarized in Table 3.5.

TABLE 3.5
Eight Sanitation Conditions Described in SSOP Documents

Sanitation Conditions	Description
Safety of water and ice	All water used in the plant is from a reliable municipal water system. The water system is designed and installed by a licensed plumbing contractor and meets current community building codes. All water faucets and fixtures are installed with antisiphoning devices.
	The ice that comes into contact with food produced is made from safe water supply, stored separately from ice used for refrigeration, and dispensed by using a scoop.
	The quality of water, steam, or ice that comes into contact with food is regularly monitored to ensure the safety.
Condition and cleanliness of food contact surfaces	Before daily processing begins, food contact surfaces are rinsed and sanitized.
	During breaks, major solids are physically removed from floors, equipment, and food contact surfaces, followed by rinsing with cold water, scrubbing using brushes with a chlorinated alkaline cleaner, and rinsing all surfaces and floors with cold water.
	At the end of daily operations, major solids are physically removed from floors, equipment, and food contact surfaces, the equipment is disassembled for adequate cleaning, and the same cleaning and sanitizing procedure used during breaks is applied again.
Prevention of cross-contamination	Good employee practices, including proper hair restraints, glove use, hand washing, personal belongings storage, eating and drinking, and boot sanitizing, are followed to avoid food contamination.
	Plant grounds are in a condition that protects against food contamination. Waste is removed from processing areas during production. The buildings are maintained in good repair. Raw product processing and cooked product processing areas are separated. Raw and cooked products are physically separated in coolers. Packaging materials are protected from contamination during storage. Cleaning and sanitizing equipment is color coded for specific plant areas.
Hand washing and toilet facilities	Toilet facilities are physically separated from processing areas, maintained in good repair, and cleaned and sanitized daily at the end of the operations.
	Hand washing/sanitizing facilities, liquid sanitizing hand soap, hand sanitizer solutions, sanitary towel service, and signs directing workers to wash their hands and gloves thoroughly are provided in processing areas and in the toilet facility.
	Hand washing and sanitizing should be applied before starting work, after each absence from the workstation, and anytime hands have become soiled or contaminated.

Protection of food from adulteration	Cleaning compounds, sanitizers, and lubricants are checked at the time of receipt to approve their use in food plants. Chemicals and lubricants are stored separately outside processing and packaging areas.
	Sanitation supervisor inspects processing and packaging equipment before operations to ensure they are in good repair with no loose or missing metal parts. No drip or condensate contaminates food or packaging materials.
Labeling, storage, and use of toxic compounds	Toxic compounds are checked at the time of receipt to ensure the information on the manufacturer's name, use instructions, and appropriate EPA approval are labeled.
	Toxic compounds are properly labeled, kept in a closed and locked cage, and stored in dry storage outside processing and packaging areas and separately from food-grade chemicals, food-grade lubricants, and packaging material storage.
	Only authorized personnel can use and store the toxic compounds during operations by following all manufacturers' instructions and recommendations.
Monitoring employee health	Workers' health conditions that might result in food contamination should be reported to their immediate supervisor.
	The workers with a potential risk are sent home or reassigned to non–food-contact jobs. Cuts or lesions on parts of the body that may make contact with food should be covered with an impermeable bandage.
Pest control	A pest management firm treats the outside of the building, and inspects the interior of the building and treats as necessary with appropriate chemicals every other month.
	Plant grounds and interior areas should be free of litter, waste, and other conditions that might attract pests. No pets are allowed in the plant, outer plant doors are kept closed, processing areas are screened with plastic curtains, and electric bug-killing devices are installed outside entrances to processing areas.

Source. Adapted from the National Seafood HACCP Alliance, Example SSOP Plan and Sanitation Control Records. Retrieved from http://nsgd.gso.uri.edu/flsgp/flsgp00001/flsgpe00001_part7.pdf.

Sufficient daily records must be maintained to document the implementation and monitoring of the SSOPs. If the SSOPs fail to prevent direct contamination or adulteration of a product, the establishment must implement corrective actions, including appropriate disposition of product, restoration of sanitary conditions, and measures to prevent recurrence. The corrective actions must also be recorded in documents.

SSOPs are the key to the successful implementation of an HACCP system. Insanitary conditions can directly cause food hazards, especially microbiological hazards. Therefore, processors of specialty food products that are subjected to the HACCP system must monitor SSOP conditions and practices during processing, and take corrective actions if sanitation conditions and practices are not met.

Ensuring Food Safety of Specialty Food Products

All food production, whether specialty or mainstream, should follow common rules and guidelines to produce wholesome products. Specialty food processors have the responsibility to ensure their food production is in compliance with applicable state or federal food safety regulations. To accomplish this, the processors need a thorough understanding of the nature of their products, the processing procedures, and the potential safety hazards associated with the food being produced. The information provides a foundation for processors to choose the most suitable food safety program in their production.

While most food safety programs are voluntary, some are mandatory by FDA/USDA, such as the HACCP system for juice, seafood, and meat products. The benefits of the food safety programs introduced in this chapter are summarized in Table 3.6. Based on the business size, the nature of the product, the potential hazards, and the pursued benefits, specialty food processors should choose the program that is most beneficial for the business to be implemented in food processing and production.

The FDA Food Safety Modernization Act (FSMA), which was signed into law in January 2011, is designed to shift food safety regulation to prevention, rather than control or investigation after a problem is found.[24] The legislation requires food manufacturers and processors to perform an analysis of food safety hazards that are reasonably likely to occur in each facility, implement controls to prevent these hazards, monitor the preventive controls to ensure that such controls are effective, and to correct them if they are not. The entire plan has to be documented, updated every 2 years or sooner if the company changes suppliers, processes, or ingredients, and shared with the

TABLE 3.6
Benefits of Different Food Safety Programs

Safety Program	Benefits
GFSI[6]	
For retailers	• Improved production standards
	• Improved information on food safety schemes
	• Exchange of best practices
	• Simplified purchasing procedures
For manufacturers	• Improved cost efficiency
	• Reduced numbers of audits
	• Clarity of food safety scheme requirements
	• Time and resources to reinvest in quality and safety
For certification bodies	• Exchange of best practices
	• Improved auditor competence and quality
	• New market opportunities
For standard owners	• Exchange of best practices
	• Greater transparency in the food industry
	• Continuous improvement
	• Market opportunities
For accreditation bodies	• Exchange of best practice
	• Knowledge sharing
	• Opportunities to work with the food industry on auditing standards
GAPs[32]	• Improve the safety and quality of food and other agricultural products
	• Reduce the risk of noncompliance with national and international regulations, standards, and guidelines regarding permitted pesticides, maximum levels of contaminants in food and nonfood agricultural products, as well as other chemical, microbiological, and physical contamination hazards
	• Promote sustainable agriculture and contribute to meeting national and international environment and social development objectives
HACCP[15]	• Provide a thorough understanding of food safety issues affecting the business and confidence in the products
	• Save business money through reduced waste, better use of manpower, and less documentation once focus is achieved
	• Maximize the benefits through focusing on the critical control points of the production process
	• Offer a cost-effective method of gaining modern management skills
	• Offer a legal defense in the event of an outbreak of food-borne disease
	• Increase trading opportunities to those companies seeking to expand their markets
BRC[17]	• Comprehensive and focused on product safety, legality, and quality
	• Clear and detailed requirements based on the adoption and implementation of a HACCP system
	• Standardized reporting format that provides information on how sites meet the requirements of the standards

(Continued)

TABLE 3.6 (CONTINUED)
Benefits of Different Food Safety Programs

Safety Program	Benefits
	• Closure of all nonconformities identified at an audit, which provides evidence in the report before certificates can be issued
	• Complementary with existing quality management systems, e.g., ISO and HACCP
ISO 22000[33]	• Organized and targeted communication among trade partners
	• Optimization of resources both internally and along the food chain
	• Improved documentation
	• Better planning and less postprocess verification
	• More efficient and dynamic control of food safety hazards
	• All control measures subjected to hazard analysis
	• Systematic management of prerequisite programs
	• Wide application due to its focus on end results
	• Valid basis for taking decisions
	• Increased due diligence
	• Control focused on what is necessary
	• Saved resources by reducing overlapping system audits
SQF[34]	• Enhance the food safety to the highest standards
	• Demonstrate the commitment to supplying high-quality, safe food products
	• Protect and enhance brand/corporate image
	• Aid in the inspection by regulatory authorities/other stakeholders
	• Enhance the confidence of consumers/customers
	• Improve new market/new customer prospects
SSOP[19]	• Assist in successful implementation of a HACCP system
	• Give clarity to the personnel, to follow steps/procedures, systematically and uniformly
	• Ensure the production of safe food
	• Prevent equipment deterioration and increase production efficiency
	• Empower employee with confidence, performance, and communication

FDA upon request.[25] If small producers do not have the expertise to identify problems themselves, they must provide valid documentation, including licenses, inspection reports, certificates, permits, credentials, certification by an appropriate agency, or other evidence of oversight, to the FDA to show the compliance of the facility with all state, local, county, or other applicable food safety laws.

The following sections give the examples of implementing specific food safety controls in the processing of some specialty food products, adapted from our previous publication of *Ensuring Food Safety in Specialty Foods Production.*[2] More detailed information for ensuring food safety and quality of specific specialty food products can be found in the different chapters of this book.

Acidified Foods

Acidified food is defined by the FDA in 21 CFR Part 114.3(b) as a low-acid food to which acids or acid foods are added to produce a product that has a finished equilibrium pH of 4.6 or below and a water activity greater than 0.85. Equilibrium pH means the final pH measured in the acidified food after all the components of the food have achieved the same acidity. The pH value of 4.6 is important because it is the limiting factor for the growth of *Clostridium botulinum*, a microorganism that is not destroyed by pasteurization or cooking temperatures below 100°C and produces a potent toxin that causes the lethal disease botulism. Examples of acidified food may include acidified artichoke hearts, marinated beets, mushrooms, and fresh-pack pickles. Acidified foods shall be manufactured, processed, and packaged to achieve a finished equilibrium pH value of 4.6 or lower within the time designated in the scheduled process and maintained in all finished foods (21 CFR Part 114.80(a)). However, some barriers exist in the preparation of acidified foods, including inadequate acid in the cover brine to overcome buffering capacity of the food, the presence of alkaline compounds from peeling or other processing aids, and the peels, waxing, piece size, or oil in the product causing a barrier to penetration of the acid. These barriers may cause the failure to achieve the final equilibrium pH of ≤4.6 and raise concerns on the growth of pathogens and production of toxins in the final products.

All acidified foods must be heat processed to destroy microorganisms that cause spoilage and to inactivate enzymes that might affect color, flavor, or texture of the product. They can be heat processed in a boiling water canner or by low-temperature pasteurization. The processing time, temperature, and procedure necessary to safely preserve the acidified foods are determined by level of acidity (pH), size of food pieces (density), and percentage salt. An FDA-recognized process authority must review the product and make the appropriate recommendations for time and temperature requirements. Temperatures higher than 85°C may break down the pectin and cause unnecessary softening of acidified foods.

Specific control measures for ensuring food safety of acidified foods may include the following:

- Acidified foods must be properly acidified to a pH below 4.6, but in practice this value is usually 4.2 or below for safety reasons.
- To ensure quick and proper acidification, the food is normally cooked or heated with the acid before being filled into the final container.

- A thermal process or heating step is required to kill all the pathogens and any other non-health-significant microorganisms that could grow during the shelf-life of the product, which must be done either by hot filling the product or by the boiling water bath process. The heating temperature and time are critical factors that must be monitored, controlled, and documented.
- The final equilibrium pH must be checked, controlled, and documented after the product has received the heating step. A pH meter with two decimal places' accuracy must be used to measure the pH if the final pH is 4.0 or above; other methods can be used, such as pH paper or a pH meter with one decimal place, if the pH is below 4.0. Containers for acidified foods should be such that a hermetic seal is obtained. Vacuum is a good indicator of a hermetic seal and helps to keep the quality of the product.
- Final products should be protected from recontamination.

Raw Milk Cheeses

Some Mexican-style Cheeses, such as queso fresco and queso cotija molido, have been responsible for outbreaks of food poisoning by several types of bacteria.[26] Mexican-style cheeses are often prepared from raw milk because pasteurization can decrease flavor and lengthen the ripening time of cheese. Due to the lack of pasteurization, raw milk Mexican-style cheese cannot be guaranteed to be free from pathogenic bacteria. During the cheese-making process, some pathogens are inactivated depending on the temperature and pH during production and ripening, yet many may survive this aging process and cause serious infectious diseases, including listeriosis, brucellosis, salmonellosis, and tuberculosis in consumers.

The USDA requires that cheeses made from unpasteurized milk must be aged for greater than 60 days at a temperature not less than 1.7°C (21 CFR Part 133.182). The aging process allows for a combination of factors, which include pH levels, salt content, and water activity to render cheeses microbiologically safe for consumption. The lactic acid formed during this aging process has been shown to inhibit the growth of pathogenic bacteria and to kill off any existing pathogens. Therefore, the storage conditions of each batch of aging cheese must be monitored by measuring room temperature and recording storage length to ensure unpasteurized milk cheese is aged for a minimum of 60 days at ≥1.7°C. If the temperature drops below 1.7°C, cheese makers must increase aging by 1 day for each day under 1.7°C. The FDA also has regulations on cheeses and related cheese products (21 CFR Part 133).

Possible control measures for ensuring food safety of Mexican cheese include the following:

- The raw milk shall be obtained from approved sources, collected and maintained in good hygienic conditions.
- If the milk is held more than 2 hours between time of receipt and setting, it shall be cooled to 7.2°C or lower until time of setting.
- Good conditions of hygiene, such as frequent cleaning of food contact surfaces, shall be maintained during production of the cheese to prevent contamination.
- The cheese must be aged for greater than 60 days at a temperature not less than 1.7°C.
- The cheese shall be refrigerated to minimize multiplication of bacteria.

The labeling requirement for raw or unpasteurized milk products varies by the states. For example, in Washington, warning labels are required on the products to establish a consumer advisory that discloses to consumers which items contain raw or unpasteurized milk and warns the consumer of the risk of consuming the product, especially by pregnant women and immunocompromised individuals. Therefore, the labeling of raw milk cheese products shall comply with the state's labeling laws or regulations.

Specialty Fruit Juice

If improperly handled, fruit juices can harbor pathogenic microorganisms and have been associated with food-borne illness outbreaks.[27] The FDA has issued regulations that mandate the application of HACCP principles to the processing of fruit and vegetable juices (21 CFR Part 120). The HACCP plans shall include control measures that will consistently produce a minimum of 5 log reduction, for a period at least as long as the shelf-life of the product when stored under normal and moderate abuse conditions, in the pertinent microorganism (21 CFR Part 120.24). For the purposes of this regulation, the "pertinent microorganism" is the most resistant microorganism of public health significance that is likely to occur in the juice, e.g., *E. coli* O157:H7.

Pasteurization is a critical control point in juice processing. The heat process used in pasteurization increases the shelf-life of juice by inactivating microorganisms and certain enzymes. To better maintain the color and flavor, flash pasteurization, also called high-temperature short-time (HTST) processing, is widely used for fruit juices. It will provide a safe product for the public, yet keep to a minimum amount the flavor degradation found in ultrapasteurized products. In flash pasteurization, the minimum temperature used is 71.5°C for a holding time between 15 and 30 seconds. To ensure the success of pasteurization, the temperature of juice needs to be continuously monitored by temperature recorder during the pasteurizing process. When monitoring indicates a deviation from the established critical limit, juice producers must segregate and hold the affected product for evaluation, destroy or divert to nonfood use, and adjust the pasteurizer (temperature or flow rate) to achieve the critical limit. The accuracy of the temperature recording device needs to be checked daily against a mercury and glass thermometer. The mercury and glass thermometer should be annually calibrated.

Possible control measures for ensuring the safety of specialty fruit juices include the following:

- A supplier guarantee must exist to specify that the shipment includes only fruit harvested, to exclude fallen fruit.
- The fruit must be rinsed and then brush washed with a sanitizer containing a minimum of 200 ppm of available chlorine for 30 seconds contact time.
- A pasteurization process with a minimum temperature of 71.5°C for 15–30 seconds is required to provide a 5 log reduction of the pertinent pathogen. The heating temperature and time are critical factors that must be monitored, controlled, and documented.

Dried Beef Jerky

Beef jerky is generally considered to be shelf-stable; i.e., it does not require refrigeration after proper processing due to the lack of moisture. However, multiple outbreaks of illness involving *E. coli* O157:H7 or *Salmonella* have been linked to beef jerky,[28] which raised the concerns on the lethality of jerky-making processes.

In 2007, the USDA Food Safety and Inspection Service (FSIS) published the *Compliance Guideline for Meat and Poultry Jerky Produced by Small and Very Small Plants* to provide updated information and guidance to small meat processors.[29] Within this guideline, the lethality treatment and drying steps are required in all processes to ensure that a safe product is produced. To have sufficient lethality, a thermal treatment combined with the 90% humidity parameter (moist cooking) must be applied in a jerky-making process to achieve a 5.0 log reduction in *E. coli* O157:H7 and *Salmonella*. After the lethality treatment, the product must be dried to meet a water activity level that will stabilize the finished product for food safety purposes.

Achieving a water activity of 0.85 or less is critical for controlling the growth of all bacterial pathogens of concern.

Critical control measures for ensuring food safety of dried beef jerky may include the following:

- A heat treatment using the time-temperature combinations provided in the lethality compliance guidelines[30] must be used in the jerky-making process to ensure sufficient lethality. The heating temperature and time are critical factors that must be monitored, controlled, and documented.
- The relative humidity must be maintained above 90% throughout the cooking or thermal heating process to meet the lethality performance standards. Humidity during heating is a critical factor that must be monitored, controlled, and documented.
- After the lethality treatment, the product must be dried to achieve a water activity level of 0.85 or less to stabilize the finished product for food safety purposes. The water activity in the final product must be checked, controlled, and documented.
- The final product must be protected from moist conditions by packaging. Gel desiccant pillows may be used to control moisture.

Tahini (Sesame Paste)

Tahini, a paste of ground sesame seeds, is a traditional food in Asia and North Africa. It is made by milling cleaned, dehulled, and roasted sesame seeds and sold fresh or dehydrated. Tahini is mainly composed of oil and protein, and its pH value ranges between 5.65 and 6.0, with an average of 5.9, while the water activity (a_w) ranges between 0.12 and 0.18, with an average of 0.16. The low a_w in tahini would not support the growth of any known food-borne microorganisms. However, low water activity does not guarantee the safety of tahini. The outbreaks of *Salmonella* infection related with the consumption of imported tahini have been reported in Australia, New Zealand, and Canada.[31] Therefore, safety controls, especially those relevant to cleanliness, personnel hygiene, and pest control, must be implemented in tahini production.

Critical control measures for ensuring food safety of tahini may include the following:

- Foreign objects must be removed from sesame seeds by using efficient sieves, magnets, and a dust suction machine.
- The water used for washing and soaking sesame seeds must be from a potable source. Total coliforms should not be detectable in 100 ml

water samples. Frequent sanitation of tanks and filters should be maintained to prevent contamination.

- A proper heat treatment (roasting at 176.7°C for 10–15 minutes) is required to eliminate disease-causing microorganisms. The heating temperature and time are critical factors that must be monitored, controlled, and documented.

Summary

It is incumbent upon all specialty food producers to provide safe food to consumers. In addition to the common food safety hazards, specialty foods may have specific safety hazards resulting from their unique processing procedures and not well-characterized ingredients. Effective safety controls must be implemented from the beginning to the end at all stages of food production to ensure that the identified potential hazards would be significantly prevented. Specialty food processors need to acquire training or education related to food ingredients, processing technologies, quality control/sanitation, packaging, food safety regulations, and specific food safety programs to generate maximum food safety assurance.

References

1. Codex Alimentarius Commission. 1997. Recommended international code of practice—General principles of food hygiene. CAC/RCP 1-1969, Rev. 3-1997. Joint FAO/WHO food standards program. http://www.fao.org/docrep/005/Y1579E/y1579e02.htm.
2. Duan, J., Y. Zhao, and M. Daeschel. 2011. *Ensuring food safety in specialty foods production*. OSU Extension publication (EM 9036).
3. Whole Foods Market. 2011. Product recalls. http://www.wholefoodsmarket.com/products/product-recalls.php.
4. Consumer Goods Forum. 2011. Global food safety initiative (GFSI). http://www.ciesnet.com/2-wwedo/2.2-programmes/2.2.foodsafety.gfsi.asp.
5. Global Food Safety Initiative. 2011. About GFSI. http://www.mygfsi.com/about-gfsi/vision-mission-and-objectives.html.
6. Global Food Safety Initiative. 2006. Global food safety initiative mission. http://www.ciesnet.com/pfiles/programmes/foodsafety/2006_GFSI_Mission_Document.pdf.
7. Global Food Safety Initiative. 2007. GFSI guidance document. 5th ed. http://www.ciesnet.com/pfiles/programmes/foodsafety/GFSI_Guidance_Document_5th%20Edition%20_September%202007.pdf.
8. Global Food Safety Initiative. 2009. The benchmark process in 10 steps. http://www.mygfsi.com/gfsifiles/GFSI_Benchmarking_Process_February_2009.pdf.

9. Food and Agriculture Organization of the United Nations. 2003. Development of a framework for Good Agricultural Practices. http://www.fao.org/docrep/meeting/006/y8704e.htm.
10. Iowa State University Extension. 2004. On-farm food safety: Guide to good agricultural practices (GAPs). http://www.extension.iastate.edu/Publications/PM1974a.pdf.
11. Global GAP. 2011. Welcome to global GAP. http://www.globalgap.org.
12. U.S. Department of Agriculture. 2009. USDA Good Agricultural Practices and Good Handling Practices audit verification checklist. http://agriculture.sc.gov/userfiles/file/Ag%20Net/audit%20checklist.pdf.
13. Goodrich, R.M., K.R. Schneider, and R.H. Schmidt. 2005. HACCP: An overview. http://edis.ifas.ufl.edu/fs122.
14. U.S. Food and Drug Administration. 1997. Hazard Analysis and Critical Control Point principles and application guidelines. http://www.fda.gov/Food/FoodSafety/HazardAnalysisCriticalControlPointsHACCP/HACCPPrinciplesApplicationGuidelines/default.htm.
15. Taylor, E. 2001. HACCP in small companies: Benefit or burden? *Food Control* 12(4): 217–22.
16. British Retail Consortium. 2011. British retail consortium. http://www.brc.org.uk.
17. BRC Global Standards. 2011. About the standards. http://www. brcglobalstandards.com/GlobalStandards/About.aspx.
18. QMI-SAI Global. 2009. BRC global standard for food safety. http://www.qmi.com/information_center/literature/brc_english_us.pdf.
19. Blanc, D. 2006. ISO 22000: From intent to implementation. *ISO Management Systems* May–June, 7–11.
20. SQF Institute. 2011. One world. One standard. http://www.sqfi.com/about-sqf/.
21. Food Safety and Inspection Service. 2010. Sanitation Standard Operating Procedures. http://origin-www.fsis.usda.gov/PDF/SSOP_module.pdf.
22. Keener, K. 2007. SSOP and GMP practices and programs. http://www.extension.purdue.edu/extmedia/FS/FS-21-W.pdf.
23. National Seafood HACCP Alliance. 2000. Example SSOP plan and sanitation control records. http://nsgd.gso.uri.edu/flsgp/flsgpe00001/flsgpe00001_part7.pdf.
24. U.S. Food and Drug Administration. 2011. FDA food safety modernization act (FSMA). http://www.fda.gov/Food/FoodSafety/FSMA.
25. PricewaterhouseCoopers LLP. 2011. FDA food safety modernization act. http://www.pwc.com/us/en/issues/food-safety-modernization-act/assets/pov-food-safety-modernization-act-vfinal.pdf.
26. Centers for Disease Control and Prevention. 2001. Outbreak of listeriosis associated with homemade Mexican-style cheese—North Carolina, October 2000–January 2001. *Morbidity and Mortality Weekly Report* 50(26): 560–62.
27. Vojdani, J.D., L.R. Beuchat, and R.V. Tauxe. 2008. Juice-associated outbreaks of human illness in the United States, 1995 through 2005. *Journal of Food Protection* 71(2): 356–64.
28. Eidson, M., C.M. Sewell, G. Graves, and R. Olson. 2000. Beef jerky gastroenteritis outbreaks. *Journal of Environmental Health* 62(6): 9–13.
29. U.S. Department of Agriculture. 2007. Quick guide on processing jerky and compliance guideline for meat and poultry jerky produced by small and very small plants. http://www.fsis.usda.gov/pdf/compliance_guideline_jerky.pdf.

30. USDA Food Safety and Inspection Service. 1999. Time-temperature tables for cooking ready-to-eat poultry products. http://www.fsis.usda.gov/OPPDE/rdad/FSISNotices/RTE_Poultry_Tables.pdf.

31. Unicomb, L.E., G. Simmons, T. Merritt, J. Gregory, C. Nicol, P. Jelfs, M. Kirk, A. Tan, R. Thomson, J. Adamopoulos, C.L. Little, A. Currie, and C.B. Dalton. 2005. Sesame seed products contaminated with *Salmonella*: Three outbreaks associated with tahini. *Epidemiology and Infection* 133: 1065–72.

32. Food and Agriculture Organization of the United Nations. 2008. Good Agricultural Practices: Introduction. http://www.fao.org/prods/GAP/index_en.htm.

33. Frost, R. 2005. ISO 22000 is first in family of food safety management system standards. *ISO Management Systems*, November–December, 16–19.

34. SGS. 2011. SQF. http://www.us.sgs.com/sqf?serviceId=10260&lobId=18866.

4

Bread and Other Baked Goods

Devin J. Rose

University of Nebraska–Lincoln
Lincoln, Nebraska

Ќ

Contents

Introduction

Today's $63 billion specialty food industry is fueled by savvy consumers that demand innovative, healthy, and convenient new food products.[1] A recent publication by the Institute of Food Technologists articulated six trends that consumers are looking for in new food products.[2] In particular, consumers want products that (1) help control major health concerns, such as body weight and diabetes; (2) contain less sodium; (3) have "clean" labels (e.g., preservative-free, all natural, organic); (4) contain nutraceutical or functional compounds; (5) are "free" of certain ingredients, such as gluten, lactose, and high-fructose corn syrup; and (6) contain niche flavors. These trends clearly express consumers' desires for healthy, delicious food products.

Baked goods offer an ideal venue for delivering products that exhibit many of these characteristics to consumers. However, it is important to consider how changes in formulation or processing may affect quality of the final product. This chapter will highlight the opportunities and challenges surrounding the creation of healthy, high-quality baked goods for the specialty foods category. Emphasis will be placed on whole wheat, multigrain, sourdough, organic, gluten-free, reduced sodium, and functional baked goods, which represent major growth categories in the baked goods market. Because yeast-leavened bread is the most important product in the baked goods category accompanied by a larger body of literature, emphasis will be placed on yeast-leavened bread; however, many of the techniques and challenges that are discussed with yeast breads can be applied to other baked goods.

Whole Wheat

With the increasing evidence supporting the use of whole grains in the diet and the recommendation that half of the grain products in the diet come from whole grain foods,[3] consumers have demanded more whole grain options; 86 whole grain products were introduced in 2007 compared to only 16 in 1997.[4] The increased demand for whole grain products doubled whole wheat flour production from 2003 to 2007[4], and production continues to climb (about 10–12% in 2011).

Bread

Whole wheat breads have traditionally been known for having lower volume, coarser texture, faster staling, and lower sensory scores than breads made from refined flours.[5,6] Therefore, much research has been devoted to improving these aspects in whole wheat bread. Food scientists have been skilled at developing and using dough conditioners, yeast nutrients, and preservatives that produce whole wheat breads with volumes, textures, and shelf stability that mirror those of white breads.[7] However, in the age of the nutritionally conscious consumer, many of these ingredients may be undesirable—particularly for the specialty bread consumer (Figure 4.1). Therefore, in the creation of specialty whole wheat breads, bakers must use other ingredients or techniques to produce acceptable breads.

The simplest ingredient to replace in a whole wheat bread formulation is high-fructose corn syrup. This is most often added because of lower cost, but may be replaced with a variety of more natural sweeteners, such as honey or

FIGURE 4.1
Ingredient Statements for Three Representative Brands of Whole Wheat Bread*

Whole Wheat Flour, Water, **High Fructose Corn Syrup**. Honey, Yeast Contains 2% or Less of each of the Following: Vegetable Oil (Soybean and/or Cottonseed Oils). Calcium Sulfate, Brown Sugar, Salt, Dough Conditioners (May Contain One or More of the Following: Mono-and Digycerides. **Ethoxylated Mono-and Diglycerides**, Ascorbic Acid, **Azodicarbonamide**, Enzymes). Yeast Nutrients (monocalcium Phosphate, Calcium Sulfate, **Ammonium Sulfate**). Corn Starch, **Calcium Propionate** (Preservative). Distilled Vinegar, Guar Gum, Sugar, Natural and Ariticial Flavor, Soy Lecithin, Soy Flour.

Whole Wheat Flour, Water, **High Fructose Corn Syrup** or Sugar, **Wheat Gluten**. Yeast, Extracts of Malted Barley and Corn Grits, Molasses. Contains 2% or Less of Less of Soybean Oil, Salt, Dough Conditioners (May Contain: **Sodium Stearoyl Lactylate, Datem**, Mono and Diglycerides, **Ethoxylated Mono and Diglycerides**, Calcium Dioxide and/or Dicalcium Phosphate). Calcium Sulfate, Yeast Nutrients (May Contain: Moncalcium Phosphate, **Ammonium Phosphate**, **Ammonium Sulfate** and/or **Ammonium Chloride**), Wheat Starch, Cornstarch, Vinegar, Soy Flour, Enzymes. **Calcium Propionate** (To Retain Freshness), Whey, Soy Lecithin.

Whole Wheat Flour, Water, **High Fructose Corn Syrup, Wheat Gluten**. Yeast, Contains 2% of: Soybean Oil, Salt, Molasses. Dough Conditioners (**Datem, Ethoxylated Mona and Diglycerides. Sodium Stearoyl Lactylate**. Mono and Diglycerides, Calcium Dioxide). Wheat Bran, Vinegar, Yeast Nutrient (Monocalium Phosphate), Calcium Sulfate, Enzymes, **Calcium Propionate** (To Retain Freshness).

* Ingredients that specialty bread consumers may not desire are in bold.

raw (turbinado) sugar. The cost associated with this replacement should not be a burden to the specialty bread baker.

Some consumers may frown upon the addition of wheat gluten to whole wheat bread formulations. Although wheat flour already contains gluten, added gluten may be perceived as being poorly digestible (see Section 4.6). Using hard wheats with strong gluten can help avoid the necessity of adding gluten. Co-proteins can also help improve the strength of bread dough. Pawar and Machewad[8] added protein concentrate from sunflower meal to determine the effects on bread quality. They found that addition of up to 10% sunflower protein concentrate improved physicochemical properties and organoleptic qualities of whole wheat bread. Soy protein isolate or soy flour has also been used to boost the protein content of whole grain breads.[9,10] Rawat et al.[9] showed that soy flour improved the softness of whole wheat bread at low levels; however, soy can impart beany flavors at high levels and also absorbs too much water during mixing, resulting in a gummy interior.[10,11]

It is difficult to remove dough conditioners, yeast nutrients, and preservatives from a whole wheat bread formulation and maintain a soft, light, and shelf-stable whole wheat bread. Dough conditioners such as mono- and diglycerides, diacetyl tartrate esters of monoglycerides (DATEMs), ethoxylated mono- and diglycerides, and sodium steroyl lactylate are added to increase dough strength and gas retention during fermentation. They can also form complexes with starch, which increases softness and reduces staling.[12]

Alkylresorcinols are compounds naturally found in rye bran and other cereals that possess amphiphilic and anionic structures similar to many dough conditioners added to bread. Andersson et al.[13] added these compounds to bread in crude and semipurified forms to determine if they could be used as more natural substitutes for synthetic dough conditioners. Unfortunately, they found that the alkylresorcinols either had no effect on bread quality (at low levels) or reduced loaf volume and crumb structure (at high levels). Alkylresourcinols were also found to reduce yeast metabolism.

Boz et al.[14] used some novel additives, such as defatted *Cephalaria syriaca* flour and rosehips, in organic whole wheat bread to improve the loaf volume. They found that a combination of 0.5% defatted *Cephalaria syriaca* flour and 2.5% rosehip increased extensiograph resistance and decreased the adhesion and stringiness of dough—parameters that are desirable for bread.

Enzymes can also help fulfill the roles of dough conditioners in baking (Figure 4.2). Food companies offer a variety of enzymes to improve the baking quality of flour (Table 4.1). Amylolytic baking enzymes partially hydrolyze amylopectin branches, which prevents retrogradation of amylopectin branches and staling. The role of xylanolytic enzymes is to improve loaf

FIGURE 4.2

Specific Volume (A) and Firmness (B) of Bread Made with Diacetyl Tartrate Esters of Monoglycerides (DATEMs; 0.4%) and Xylanase (100 ppm, Bel'Ase B210, Beldem-Purato)

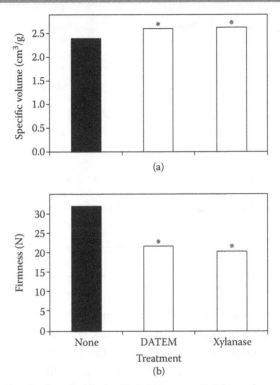

(a)

(b)

Source. Based on data from Grausbruber, H., S. Meisenterger, R. Schoentechner, and J. Vollmann, *Acta Alimentaria,* 37, 2008, pp. 379–390.

* Indicates data is significantly different from the control.

volume. Baking xylanases selectively hydrolyze water-unextractable arabi-noxylans in wheat flour. This serves to diminish the detrimental effects of water-unextractable arabinoxylans on loaf volume and texture (Figure 4.2). Because water-unextractable arabinoxylans are so much more concentrated in whole wheat bread than in traditional white breads, xylanases may affect whole wheat bread volume to an even greater extent than white bread. Transglutaminase cross-links proteins and thus strengthens protein networks. This enzyme is much less common in baking enzyme preparations than amy-lases and xylanases. Tranglutaminase may enhance immunogenicity in indi-viduals sensitive to gluten; therefore, its use may not be recommended.[16,17]

TABLE 4.1
Some Examples of Commercial Baking Enzyme Preparations

Enzyme(s)	Function	Usage[a]	Trade Name	Supplier
Maltogenic amylase	Prevent staling		Novamyl 10.000 BG	Novozymes
Fungal α-amylase + xylanase	Prevent staling/improve loaf volume		Fungamyl Super MA	Novozymes
Xylanase	Improve loaf volume	100 ppm	Grindamyl PowerBake 900	Danisco A/S
Xylanase	Improve loaf volume	100 ppm	Bel'Ase B 210	Beldem-Puratos
α-Amylase	Prevent staling	20 ppm	Grindamyl A 1000	Danisco A/S
α-Amylase	Prevent staling	20 ppm	Bel'Ase A 75	Beldem-Puratos
α-Amylase + xylanase	Prevent staling/improve loaf volume	100 ppm	Bel'Ase 2000B	Beldem-Puratos
α-Amylase + xylanase + transglutaminase	Prevent staling/improve loaf volume	100 ppm	Veron CLX	AB Enzymes

[a]   Flour weight.

In addition to dough conditioners, specialty bread consumers may frown upon the use of ammonium and other salts used as yeast nutrients. Germinated (malted) flours are often added to bread as yeast food and are more label-friendly. Interestingly, Seguchi et al.[18] added bran from germinated wheat to bread at 10% (flour weight) and found that it improved the loaf volume. They attributed this to enzyme activities expressed in the germinated bran, namely, xylanase and α-amylase. This may be used in a whole grain setting to improve bread quality.

One aspect of whole wheat bread baking that must not be overlooked is the flour type. Most wheat produced in the United States is red wheat; therefore, most whole wheat flour is from red wheat. As whole wheat flour demand increases, white whole wheat flours are becoming more available. These flours are nutritionally comparable to red wheat flours, but are generally more desirable due to their lighter color and reduced bitterness.[19]

Because most mills are set up for traditional refined flour production, most commercial whole wheat flours are produced by recombining the flour, bran, and shorts fractions obtained from roller milling in the approximate proportions to make whole wheat flour. Much less common are whole wheat flours produced from milling of the whole grain without separation using stone mills. Kihlberg et al.[20] evaluated roller-milled and stone-milled wheat for baking and sensory quality of bread produced from the flours. Whole wheat breads produced from roller-milled wheat were described by a trained sensory panel as being sweet, juicy, and compact. Stone-milled wheat breads were described as salty, roasted, and nonuniform. Importantly, milling technique had a greater effect on final bread quality than baking technique.

Bran particles coming off a roller mill can be quite coarse. Therefore, millers often remill the bran before recombining it with flour and shorts to make whole wheat flour. Depending on the technique used in this second milling operation, widely different bran particle sizes can be obtained. This affects bread-making quality of the flour. In general, very finely milled bran particles reduce loaf volume.[6,20,21] Fine grinding increases the surface area of the bran particles and increases interaction with gluten. This results in reduced gluten strength and gluten network formation.[22] Conversely, coarsely milled bran contributes to a rough crust appearance and gritty texture.[6] These results suggest that a moderate bran particle size is most desirable for whole wheat bread baking.

Other Baked Goods

Few studies have been conducted on whole wheat baking of items other than bread. Consumer acceptance of whole wheat muffins made with red or white wheat has been evaluated. Muffins made with red wheat were perceived as more healthy than those made with white wheat. When consumers were provided with nutritional information for the muffins, sensory scores increased more for the white wheat muffins than for the red wheat muffins.[23]

As with whole wheat bread, whole wheat flour particle size is important in baking properties of other baked goods. Gaines and Donelson[24] evaluated 69 soft wheat cultivars for whole wheat cookie baking quality. They found that cookie diameter increased as flour particle size increased. A number of studies have addressed adding bran to baked goods as a means of improving nutrition. While this is not strictly whole wheat, it can help illustrate the effects of bran particle size on product quality. Gomez et al.[25] substituted up to 36% of the flour in layer cakes with wheat bran of different particle sizes (50, 80, 250 μm). They found that large particle sizes increased firmness, chewiness, and yellowness in cakes and reduced sensory scores to the greatest degree. As Noort et al.[22] illustrated, finely milled bran particles interact with gluten and reduce strength. This suggests that while finely milled bran particles do contribute some negative aspects to product quality, such as darkening, reduced volume, and grittiness, they may help to "shorten" the dough in products that require low gluten development. Thus, in baked goods that do not rely on gluten for proper texture, using finely milled whole wheat flour may be the best approach.

Multigrain/Seed

Much research has gone into whole wheat research in baked goods; however, other grains and seeds may be used alone or in combination with wheat flour in baked goods. Oftentimes the addition of these ingredients produces an inferior product (e.g., references 26–28); however, sometimes these other grains or seeds can impart intriguing new flavors, enhance nutritional profile, or even improve quality.

Bread

Bread can take on sweet, savory, and other flavors as a result of the addition of nonwheat grains or seeds. For instance, Indrani et al.[29] supplemented bread with up to 7.5% (flour weight) fenugreek seed powder in Indian flat bread. Fenugreek seeds have a maple aroma and flavor and are often used to flavor curries. They found that 5% addition created an acceptable bread with a distinct fenugreek flavor. Lin et al.[30] baked bread containing 15% (flour weight) husked and unhusked buckwheat (*Fagopyrium esculentum* Moench) flour and analyzed free amino acids, 5'-nucleotides, and volatile flavor compounds to quantify the effects of buckwheat flour on bread flavor perception. They found that both buckwheat flours increased umami flavors, an increasingly popular flavor in the specialty food arena. Husked buckwheat also gave a honey-sweet aroma, while the unhusked buckwheat bread imparted a milky, roasted aroma. Bodroza-Solarov et al.[31] prepared breads containing up to 20% (flour weight) amaranth (*Amaranthus cruentus*) flour. Although the loaf volume and crumb elasticity decreased and crumb hardness increased as a function of amaranth flour level, a sensory panel rated the breads containing 10–15% amaranth flour as high as the control breads for flavor. Panelists described the amaranth-supplemented breads as pleasant, aromatic, nutty, and roasted.

Not only can exotic ingredients impart pleasing unique flavors, but they oftentimes carry with them added nutritional benefits. For instance, Sullivan et al.[32] substituted pearled barley flour for wheat flour in bread. They found that up to 50% of the wheat flour could be replaced with millet flour without reducing sensory properties, and the resulting bread was higher in dietary fiber and β-glucan. Whole and ground flaxseed was added at substantial levels (36% flour weight) to bread.[33] Flaxseed is a good source of essential ω-3 fatty acids. During 6 days of storage, no off-odors from lipid oxidation were detected in the flaxseed breads. Moreover, the flaxseed bread retained moisture and softness during storage better than the control bread. Losso et al.[34] added fenugreek flour to bread at 5% (flour weight) and showed that this reduced the postprandial insulin response.

Other Baked Goods

Baked goods other than bread have been commonly studied for the effects of nonwheat flours on product quality. Sharma and Chauhan[35] added fenugreek flour to cookies containing rice bran. They found that the fenugreek flour sensory characteristics and overall cookie quality score compared with those of cookies supplemented with rice bran alone. Gomez et al.[25] replaced all of the wheat flour in cakes with sifted rye, barley, triticale (wheat-rye cross), or tritordeum (wheat-barley cross) flour. Whole grain barley flour cakes and whole wheat cakes received similar sensory scores.

In some cases, replacing wheat flour with another flour can improve organoleptic properties. High oleic sunflower seed meal was used in cookies and improved sensory scores significantly when replacing 10 or 30% of wheat flour.[36] Filipcev et al.[37] substituted 30% of the wheat flour with buckwheat flour in biscuits. The buckwheat biscuits received higher sensory scores for softness and fracturability than the control containing 100% wheat flour.

An improved nutritional profile as a result of adding nonwheat flour components to baked goods has also been demonstrated. Filipcev et al.[37] showed that buckwheat flour biscuits contained enhanced levels of rutin and quercetin, flavonoids that are associated with reduction of oxidative events. Hooda and Jood[38] added fenugreek flour to whole wheat biscuits and found that protein, lysine, dietary fiber, calcium, and iron were increased when 10% of the wheat flour was replaced with fenugreek flour. Moraes et al.[39] formulated cakes containing up to 45% flaxseed flour as a replacement for wheat flour. The formulations made with up to 30% flaxseed flour showed good sensory acceptance, and contained 6.92 g dietary fiber and 2.5 g linolenic acid per 100 g serving. Handa et al.[40] supplemented cookies with 10–30% millet or sorghum and replaced up to 60% of the sugar with fructooligosaccharides. The most acceptable cookie was produced with 20% millet and 30% sorghum replacing the wheat flour and 60% of the sugar replaced with fructooligosaccharides. The resulting cookies contained 5.2 g dietary fiber per 30 g serving and 11% fewer calories than the control cookie.

Sourdough

Fermentation, or sourdough, is an ancient method for producing baked goods that takes advantage of the metabolic activities of naturally occurring or specifically designed yeasts and lactic acid bacteria to produce characteristic flavors, aromas, and textures. Sourdough is most often associated with bread, but has been used in a few instances in other baked goods.

Bread

Yeasts and lactic acid bacteria produce acids, carbon dioxide, ethanol, and enzymes during sourdough fermentation that not only impart characteristic sourdough flavors to bread,[41] but also improve loaf volume,[42,43] texture,[44] and nutritional quality.[45] A number of researchers have shown improved loaf volume of whole grain or bran-enriched breads after fermentation.[46] One explanation for this effect has been that acidification of the dough by lactic acid bacteria leads to an increased capacity of gluten to retain carbon dioxide.[47]

As mentioned, consumers are on the lookout for products that help with control of diabetes. Whole grain breads exhibit similarly high glycemic response curves as traditional white breads.[48,49] However, sourdough whole wheat bread results in a significantly reduced postprandial glucose response compared with straight-dough white and whole wheat breads.[49,50] It is likely that this phenomenon is a result of hydrolysis and solubilization of nonstarch components in the flour, such as protein or dietary fiber,[49,50] which may interact with starch in the small intestine and delay glucose uptake.

Sourdough fermentation can also improve nutritional quality of breads by improving mineral bioavailability. Whole wheat bread is a good source of many trace minerals, including magnesium and iron. However, phytic acid can chelate these metals and prevent their absorption. During fermentation, phytases naturally present in flour can hydrolyze the phosphate groups from phytic acid and, together with acid production by the microorganisms, minerals are released from their bound state and able to be absorbed during digestion.[51,52]

Finally, sourdough fermentation may improve the shelf-life of bread. As mentioned, bacterial metabolism during sourdough fermentation produces acids, alcohol, and enzymes, which may improve shelf-life by preventing mold growth or preventing staling. Specially designed starter cultures may further improve shelf-life.[53,54] *Lactobacillus plantarum* FST 1.7 retarded the growth of *Fusarium culmorum* and *Fusarium graminearum*, two fungi found on bread, compared with fermentation with *Lactobacillus sanfranciscensis*, a common sourdough starter culture.[55] Zhang et al.[56] prepared an experimental sourdough starter containing *Lactobacillus buchneri* and *Lactobacillus diolivorans*. The cofermentation of these two lactobacilli produces propionic acid, the major preservative used in bread. They found that 20% experimental sourdough starter in the formulation inhibited the growth of *Aspergillus clavatus*, *Cladosporium* spp., and *Mortierella* spp. for more than 12 days and *Penicillium roquefortii* for 2 days compared to a control bread containing added calcium propionate. This sourdough was also more effective at inhibiting mold growth than traditional sourdough bread prepared with additional acetate.

Other Baked Goods

Although sourdough is usually associated with bread, fermentation is an important part of the production of a few other baked goods. Soda crackers in particular are fermented for up to 24 hours prior to baking. Acids produced by the lactic acid bacteria during the fermentation process lower the pH of the dough and allow proteases to become active. These proteases partially hydrolyze the gluten proteins and soften the dough.[57] This process may be useful in the production of other baked goods that require soft doughs.

Fermentation may also provide enhanced nutrition in baked goods. As has already been discussed, fermentation can reduce the glycemic index and improve mineral bioavailability of baked goods. Additionally, bacteria and yeasts that are elaborated during fermentation can produce bioactive compounds such as peptides, γ-amino butyric acid, conjugated linoleic acid, and antioxidants,[58,59] and vitamins such as thiamin, riboflavin, niacin, vitamin B-6, and folate.[58,60]

Fermentation may also allow for the production of new and exciting flavors and aromas in baked goods, such as muffins and biscuits, in much the same way that it modifies these parameters in bread. Using alternative grains or seeds may further enhance these effects.

Organic

In 1990, the U.S. Congress passed the Organic Foods Production Act (OFPA). This act required the U.S. Department of Agriculture to develop standards for organically produced products (Code of Federal Regulations (CFR) Title 7, Part 205, "National Organic Program"). Organic wheat production requires exclusion of synthetic pesticides and fertilizers and genetically modified plants. Many consumers are looking for organically grown products out of health concerns.

Bread

Literature suggests that organically grown wheat may not possess the same functional properties of conventionally grown wheat, even of the same variety. For instance, organic wheat produces poorer quality bread in terms of loaf volume and texture than conventional wheat.[20,61–64] Krejcirova et al.[65] suggested that this may be due to lower percentage of high molecular weight glutenins, proteins that are important for elasticity of dough.

In sensory studies comparing bread made from organic and conventional wheat, few differences have been identified.[20,63,66] For instance, in a study comparing three organic wheats with three conventional wheats, panelists described two of the conventional breads as having the darker crust color, more of a cereals aroma, flavor, and aftertaste, and more saltiness, while the conventional wheats and one of the organic wheats were described as having higher wheat aroma and sweetness of the crumb, and roasted cereals flavor and crispiness of the crust.[20] Annett et al.[63] found no significant differences for flavor, aroma, and color in paired organic and conventional wheats. Therefore, differences in sensory attributes between organic and conventional breads may simply be due to changes in panelist perception as a result of more compact loaf volumes.[20,63]

These data suggest that variety selection is very important for producing organic whole wheat bread with acceptable loaf volume. Much research is currently underway regarding development of new wheat varieties specifically designed for organic production. In the meantime, strategies to improve the loaf volume and baking quality of breads and other baked goods made from organic wheat are similar to those strategies employed for conventional wheats (discussed above), keeping in mind that consumers of specialty and organic foods are looking for "cleaner" labels than typical consumers.

Other Baked Goods

Little research on the use of organic vs. conventional wheat or other grains in baked goods other than bread has been reported in the literature. Despite this, organic pretzels, cookies, crackers, muffins, and other baked goods can be readily found in many supermarkets and specialty food stores. It is likely that differences in functionality among conventional and organic grains in non-yeast-baked good applications are similar to those described above.

Gluten-Free

The prevalence of celiac disease, an autoimmune disease triggered by the consumption of prolamines from wheat, rye, barley, and related cereals, has hovered around 1% of the population,[67] although recent research suggests that the prevalence is increasing.[68] As the disease has become more mainstream, the demand for gluten-free products has increased dramatically.[69] In 2009, an estimated 13.2% of the population was on the lookout for gluten-free foods,[70] and the market has increased from $580 million in 2004 to $1.56 billion in 2008.[69]

These data indicate that far more consumers are interested in gluten-free products than there are people with celiac disease. Nonceliac consumers that are in the market for gluten-free foods cite digestive health, nutritional value, and help in losing weight as the top reasons for purchasing these products.[70] Ironically, traditional gluten-free products have been mainly starch or refined rice flour based, and therefore, in addition to being low in dietary fiber, lacked nutrients such as folic acid and iron that are added to refined, wheat-based foods.[71]

Bread

Yeast-leavened bread is the ultimate challenge for the gluten-free baker. Gluten-free breads have been characterized by poor texture, crust color, shelf-life, and flavor,[72] which results from the absence of gluten, the main protein complex in wheat, which is responsible for the elasticity, strength, and gas-trapping ability of dough and the texture, volume, and appearance of bread. The viscous and gas-trapping properties of gluten have traditionally been mimicked by polysaccharide gums in gluten-free products. These gums can improve dough development and gas retention by increasing dough viscosity. Data compiled from several studies (Table 4.2) suggest that cellulose derivatives such as hydroxypropylmethylcellulose, carboxymethylcellulose, and methylcellulose may produce the most acceptable bread when used at about 2% in flour-starch-based formulations. In addition to their viscous properties, these gums in particular contain partial hydrophobic characteristics that may contribute to increased interfacial activity during dough proofing

TABLE 4.2
Gluten-Free Breads Containing Polysaccharide Gums

Base	Co-Protein	Polysaccharide(s)	Outcome(s)	Reference
White rice flour, tapioca flour, potato starch	Egg (liquid, 17.4% FSB[a]), nonfat dry milk (4% FSB)	Xanthan (1.5%), HPMC[b] (0.5%)	Gluten-free bread retained moistness and freshness better than a standard wheat bread over 120 h of storage	73
Potato starch, corn starch (corn flour also included in guar: pectin bread)	None	Pectin (5% FSB), guar (3.6% FSB), pectin + guar (1.75% each FSB)	Breads with guar were superior to breads with pectin; pectin + guar bread was most desirable	74
White rice flour with 10% defatted rice bran	None	HPMC (2.7% FSB)	Rice bread had lower specific volume, harder texture, and was more susceptible to retrogradation than whole wheat bread	75
White rice flour, corn starch	Sodium caseinate (amount not disclosed)	Pectin, CMC[c], agarose, xanthan, β-glucan (1 and 2% rice flour basis)	Highest sensory scores were obtained with CMC at 2%	76
White rice flour, potato starch	Skim milk powder (10% FSB)	HPMC (0.5–2.5% FSB)	2.2% HPMC and 79% water (FSB) gave best combination of high specific volume and low crumb firmness	77
Brown rice flour, potato starch, corn starch, buckwheat flour	Soy flour (14.3%), skim milk powder (43% FSB)	Xanthan (1.5%), xanthan (1% FSB) + konjac (1.7% FSB)	Breads containing skim milk powder produced the best bread	78
Rice flour, corn starch, corn fiber	None	HPMC (1.5% FSB)	Optimum loaf specific volume, crumb firmness, and crumb L value were obtained with 6.5% corn fiber and 102.5% water (FSB)	79
Corn starch	Zein (25% FSB)	HPMC (2.5% FSB)	HPMC improved crumb grain and stabilized gas cells, improved specific volume, and produced a rounded top	80

TABLE 4.2 (CONTINUED)
Gluten-Free Breads Containing Polysaccharide Gums

Base	Co-Protein	Polysaccharide(s)	Outcome(s)	Reference
Long-grain rice flour	None	HPMC (2.7–8.2% FSB)	HPMC (5.5% FSB) gave the best crumb structure and baking properties compared with other treatments	81
Pregelatinized rice flour, pregelatinized corn starch, corn flour	Egg albumin (dried, 1.37–4.63% FSB)	Methylcellulose, gum arabic (1.37–4.63% FSB)	Breads with the most acceptable frequency of cracks, separation of layers, rollability, tearing quality, hardness, adhesiveness, and cohesiveness were obtained with 3% gum arabic, 2.10–4.12% methylcellulose, and 2.18–4.10% egg albumin	82
White rice flour, buckwheat flour		Xanthan and propylene glycol aglinate (0.5–1.5% FSB)	Propylene glycol alginate breads had higher specific volume and lower crumb firmness than xanthan breads	83

a FSB = flour-starch basis = weight of flour + starch in formulation.
b HPMC = hydroxypropylmethylcellulose.
c CMC = carboxymethylcellulose.

and gel-forming properties during baking.[77,84] Moreover, cellulose derivatives may interact with starch granules by increasing their gelatinization temperature and allowing them to adhere to one another.[81,85]

Although bread quality is substantially improved with the addition of polysaccharide gums, the textural and staling properties of these gluten-free breads are still inferior to wheat bread.[75,82] Today new strategies aimed at improving both nutrition and quality of gluten-free foods are being developed. One strategy that is being explored is the use of whole grain flours from less common grains and pseudocereals. For instance, Alvarez-Jubete et al.[86] replaced the potato starch with amaranth, quinoa, or buckwheat flours in a gluten-free bread formulation. They found that the pseudocereals improved both volume and texture of the breads compared with the gluten-free control. They attributed this to naturally occurring emulsifiers in these flours. Since they were adding wholemeal flours, the crumb color darkened, but protein, dietary fiber, and minerals substantially increased over the potato starch-containing control.[87] No significant differences in sensory scores among any of the breads tested were discovered.

Others have also used amaranth in gluten-free applications. Mariotti et al.[88] supplemented amaranth dough with a yellow pea isolate (Pisane F9, Cosucra, Belgium) to increase protein content and psyllium husk to increase viscosity. They reported that the psyllium husk enhanced the workability of the gluten-free doughs due to the viscous and film-forming properties that it provided. De la Barca et al.[89] used a novel technique to improve amaranth functionality in bread making without using polysaccharide gums. They incorporated popped amaranth flour into rice flour–corn starch-based doughs at different percentages. The specific volume of bread increased proportionally in response to the amount of popped amaranth flour in the dough.

Oats have traditionally been shunned by people with celiac disease, but research has shown that many of these people can tolerate oats.[90] Huttner et al.[91] prepared gluten-free bread using three whole oat flours with different compositions. They found that oat flour with coarse particle size, minimal starch damage, and low protein content made the most acceptable bread in terms of loaf volume and crumb hardness.

Malting and fermentation are also new strategies to improve the quality of gluten-free breads. Fermentation of nongluten grains in the production of gluten-free breads has the potential to reduce staling due to amylase activity and improve nutrition due to phytase activity.[92] Numerous flavor compounds are also generated during fermentation that can contribute to unique and pleasing aromas and flavors.[92]

Alternatively, enzymes can be used directly in the production of gluten-free breads. Transglutaminase has already been mentioned as an enzyme that is used to enhance mixing tolerance of wheat flour breads. This enzyme may help cross-link nongluten proteins and improve cell structure and strength of gluten-free breads,[93] although this may change the innumogenicity of cereal proteins and may pose problems in patients with celiac disease.[16,17] Other enzymes such as proteases and glucose oxidase have shown some promise in gluten-free applications, although they show differing effects depending on flour type.[95]

Other Baked Goods

While the challenges involved in producing gluten-free, yeast-leavened bread are more formidable, converting non-yeast-leavened baked goods to gluten-free can impose certain challenges of its own. Many of the strategies employed for improving gluten-free bread quality can be applied to other baked goods, including the addition of gums, emulsifiers, and enzymes.[95,96] De la Barca et al.[89] used popping as a means of improving the functional properties of amaranth-rice flour cookies. They found that amaranth flour decreased cookie diameter. This was partially overcome by using popped amaranth flour, which improved cookie spread and reduced hardness.

Improving the nutrition of starch and white rice flour-baked goods has also been an active area of research. Gambus et al.[97] replaced potato starch with wholemeal amaranth, linseed, or buckwheat in cookies and cakes. They did not find significant differences in sensory scores for the supplemented cakes and cookies compared with their controls; however, the protein, dietary fiber, and nutrient content was substantially enhanced by replacing the starch with the wholemeal flours. Sedej et al.[98] produced wholemeal and refined buckwheat crackers and compared them to whole wheat and refined wheat crackers. The refined buckwheat flour crackers received significantly higher sensory scores for appearance, structure, break, firmness, and chewiness than the refined wheat crackers; the wholemeal buckwheat and whole wheat crackers performed equally well in the sensory panel. Both refined and wholemeal buckwheat crackers also contained more antioxidants, including α-tocopherol, and dietary fiber than their controls. Crackers produced from bean flours also have potential to produce a highly desirable product with enhanced nutrition.[99]

Reduced Sodium

Although often thought of a "bad for you," sodium is an essential nutrient. Along with potassium and chloride, it is an important electrolyte in the body and helps maintain osmotic and electrical balance in the body. However, intake of sodium in the diet is far beyond what is required for electrolytic function: mean sodium intake for the U.S. population in 2005–2006 was 3,436 mg/d—more than double the recent recommendations set forth in the *Dietary Guidelines for Americans, 2010* and the Dietary Reference Intakes of 1,500 mg/d.[3,100] In general, high sodium intake contributes to high blood pressure; thus the benefits of reducing sodium in the diet include less chance of stroke and cardiovascular disease events.[3]

Many consumers believe the majority of the sodium in their diet comes from processed or packaged foods; they are not aware that foods that do not necessarily taste salty can contribute significantly to the amount of sodium in their overall diet through repetitious eating.[101] Indeed, yeast breads contribute the largest percentage of sodium intake to the diet, while other baked goods, such as cakes, cookies, quick breads, and crackers, also contribute substantially.[102]

The major challenge in producing low-sodium baked goods is retaining the proper sensory profile. Heidolph et al.[103] describes three approaches to reducing sodium in foods: (1) removing all or part of the salt, (2) replacing the sodium with another ingredient, and (3) enhancing the flavor of the reduced sodium food.

Bread

In yeast breads, salt (sodium chloride) is the major contributor to the sodium content. Salt plays two functional roles in bread making: (1) it increases the osmotic pressure of the dough and (2) it partially neutralizes electrostatic repulsion among gluten proteins. The combined effects of these two functions have major impacts on bread quality. For example, the increased osmotic pressure leads to reduction in yeast metabolic rate. Combined with the strengthening effect of gluten proteins, gas bubbles expand to maximum size without rupture. Bread made without salt has abnormally large bubbles, coupled with bubbles that have collapsed due to poor gluten strength, resulting in poor loaf volume. Salt also increases dough mixing time and improves mixing tolerance through its interaction with gluten proteins. Salt also contributes to the sensory properties of bread. Not only is bread made with salt perceived as more salty, but trained panelists also perceive breads made with salt as less yeast-like, less musty, and less floury/watery, attributes that are usually undesirable in bread.[103]

The functional benefits of salt on bread dough are maximized at 0.5–2% salt (flour weight) in the formulation. Above 2% salt, crumb firmness increases, staling rate increases, and loaf volume decreases. Most bread formulations contain 1–2% salt and are therefore within the range of maximum benefit; however, lower sodium breads could possibly be made with only 0.5% salt without sacrificing loaf volume or texture.[103] Unfortunately, these breads may be perceived as bland. Adding other flavors such as herbs or spices in certain types of bread may help overcome this perception. In unflavored bread, however, this may not be practical. Noort et al.[104] used a novel approach to removing sodium in bread. They produced bread doughs with varying concentrations of salt. The doughs were then sheeted and stacked in alternating fashions to create composite doughs with varying sodium contents, but with heterogeneous distribution of salt within the loaf. They found that perceived saltiness could be maintained with a 28% reduction in salt content compared with a control loaf.

Replacing the sodium in bread can also be accomplished by using other salts. Most commonly, potassium chloride is used; however, when used as a complete replacement, potassium imparts a bitter, metallic flavor. In general, up to 30% replacement of sodium chloride with potassium chloride is acceptable in bread.[105-107] Other salts may be used to replace or reduce sodium chloride. These include magnesium chloride, calcium chloride, potassium lactate, and magnesium sulfate.

Other Baked Goods

Non-yeast-leavened baked goods contain appreciable amounts of sodium from salt and from chemical leavening agents, preservatives, and other salts (Table 4.3). The sodium portion of these salts usually does not contribute to product functionality as does sodium chloride in bread; the sodium can contribute partially to flavor, but exhibits minimal effect on product quality.

Many alternatives are available as replacements for the sodium salts used as chemical leavening agents. Some of these rely on replacement of sodium with potassium, such as potassium bicarbonate and potassium aluminum phosphate. With these salts, one must be mindful of the bitterness that potassium imparts at high levels. There are also calcium salts such as dicalcium

TABLE 4.3
Ingredients Added to Baked Goods that Contribute Sodium to the Product

Ingredient	Purpose	Sodium (%)	Usage
Disodium ethylenediaminetetraacetate	Preservative	12	Up to 315 ppm[a]
Sodium aluminum phosphate	Leavening	3	
Sodium aluminum sulfate	Leavening	5	
Sodium benzoate	Preservative	16	Up to 0.1%[a]
Sodium bicarbonate	Leavening	27	
Sodium chloride	Flavor/functionality	40	0.5–2%[b]
Sodium propionate	Preservative	24	0.1–0.75%[b]
Sodium pyrophoshpate	Leavening	10	
Sodium steroyl lactylate	Dough conditioner	5	Up to 0.5%[a]

[a] Legal limit (percent in final product): disodium ethylenediaminetetraacetate, 21 Code of Federal Regulations (CFR) 172.135; sodium benzoate, 21 CFR 184.1733; sodium steroyl lactylate, 21 CFR 172.846.
[b] Typical usage in baked goods (flour weight).

phosphate and calcium acid pyrophosphate that may replace sodium-based acids in baking powders. When replacing a sodium salt with another metal salt for chemical leavening, it is important to recognize that the replacement salt may differ in its release of carbon dioxide.

Sodium content of baked goods with a major portion of the sodium on the outside or surface of the product, such as crackers and pretzels, may also be reduced without affecting saltiness. In this case, reducing the granulation size salt or changing salt crystal structure can maintain saltiness perception while being used at lower concentrations.[103] Alternatively, improving the flavor of low-sodium foods may be achieved not by replacement but by enhancement. For instance, in baked goods where more savory flavors are appropriate, certain flavor enhancers such as yeast extract, amino acids, ribonucleotides, and organic acids may be used.

Functional Baked Goods

Often, consumers are on the lookout for specialty items that contain specific bioactive components that may reduce the risk of specific chronic diseases or beneficially affect health beyond what is contributed by basic nutrients. These foods are termed functional foods.[108] Sometimes these ingredients are added to meet a particular health claim (Table 4.4), and other times they are added as supplements. Some examples of functional baked goods are shown in Table 4.5.

Bread

Bread is an important source of dietary fiber in the U.S. diet; however, dietary fiber intake in the United States is far below recommendations.[3] Whole grain and many multigrain breads are naturally high in dietary fiber; however, there are many cases where consumers may desire products that contain dietary fibers that target specific health-related issues, such as heart disease or digestive health.

Products containing mixed linkage (1,3), (1,4)-β-glucan (β-glucan) from oats or barley may carry a U.S. Food and Drug Administration (FDA)-approved health claim indicating that the product may help reduce the risk of heart disease (Table 4.4). Many high β-glucan products are available as food ingredients.[133] Some of these products are flours produced from high β-glucan varieties of oats or barley or by special milling processes that increase the β-glucan content of the final flour. Other products are more

TABLE 4.4

U.S. Food and Drug Administration-Approved Health Claims

Claim	Example Statement	Requirements[a]	Reference[b]
Calcium, vitamin D, and osteoporosis	Adequate calcium (or vitamin D) throughout life, as part of a well-balanced diet, may reduce the risk of osteoporosis.	Must be a "good source" of calcium or vitamin D; cannot contain more phosphorus than calcium	21 CFR 101.72
Dietary lipids and cancer	Development of cancer depends on many factors. A diet low in total fat may reduce the risk of some cancers.	Must meet requirements for "low fat"	21 CFR 101.73
Dietary saturated fat and cholesterol and risk of coronary heart disease	While many factors affect heart disease, diets low in saturated fat and cholesterol may reduce the risk of this disease.	Must meet requirements for "low saturated fat," "low cholesterol," and "low fat"	21 CFR 101.75
Dietary noncariogenic carbohydrate sweeteners and dental caries	Frequent eating of foods high in sugars and starches as between-meal snacks can promote tooth decay. [Name of sweetener], the sweetener in this food, unlike other sugars, may reduce the risk of dental caries.	Eligible sweeteners: xylitol, sorbitol, mannitol, maltitol, isomalt, lactitol, hydrogenated starch hydrolysates, hydrogenated glucose syrups, erythritol, D-tagatose, isomaltulose, and sucralose	21 CFR 101.80
Fiber-containing grain products, fruits, and vegetables and cancer	Low-fat diets rich in fiber-containing grain products, fruits, and vegetables may reduce the risk of some types of cancer, a disease associated with many factors.	Must contain a grain product, fruit, or vegetable and meet requirements for "low fat" and "good source" of fiber.	21 CFR 101.76
Folate and neural tube defects	Adequate folate in healthful diets may reduce a woman's risk of having a child with a brain or spinal cord birth defect.	Must meet requirements for "good source" of folate or folic acid; must not contain >100% DV of retinol/preformed vitamin A or vitamin D	21 CFR 101.79

(Continued)

TABLE 4.4 (CONTINUED)
U.S. Food and Drug Administration-Approved Health Claims

Claim	Example Statement	Requirements[a]	Reference[b]
Fruits and vegetables and cancer	Low-fat diets rich in fruits and vegetables (foods that are low in fat and may contain dietary fiber, vitamin A, and vitamin C) may reduce the risk of some types of cancer, a disease associated with many factors. This food is high in vitamins A and C, and it is a good source of dietary fiber.	Must meet requirements for "low fat" and, without fortification, "good source" of vitamin A, C, or dietary fiber	21 CFR 101.78
Fruits, vegetables, and grain products that contain fiber, particularly soluble fiber, and risk of coronary heart disease	Diets low in saturated fat and cholesterol and rich in fruits, vegetables, and grain products that contain some types of dietary fiber, particularly soluble fiber, may reduce the risk of heart disease, a disease associated with many factors.	Must contain a grain, fruit, or vegetable and meet requirements for "low saturated fat," "low cholesterol," and "low fat" and contain, without fortification, >0.6g soluble dietary fiber/serving	21 CFR 101.77
Sodium and hypertension	Diets low in sodium may reduce the risk of high blood pressure, a disease associated with many factors.	Must meet requirements for "low sodium"	21 CFR 101.74
Soluble fiber from certain foods and risk of coronary heart disease	Soluble fiber from foods such as [name of soluble fiber source], as part of a diet low in saturated fat and cholesterol, may reduce the risk of heart disease. A serving of [name of food] supplies ____ g of the [grams of the soluble fiber necessary] per day to have this effect.	Must contain at least 0.75 g soluble fiber from oat or barley or 1.7 g from psyllium per reference amount and meet requirements for "low saturated fat," "low cholesterol," and "low fat"[c]	21 CFR 101.81

TABLE 4.4 (CONTINUED)
U.S. Food and Drug Administration-Approved Health Claims

Claim	Example Statement	Requirements[a]	Reference[b]
Soy protein and risk of coronary heart disease (CHD)	25 g of soy protein a day, as part of a diet low in saturated fat and cholesterol, may reduce the risk of heart disease. A serving of [name of food] supplies __ g of soy protein.	Must contain at least 6.25 g soy protein per reference amount and meet requirements for "low saturated fat," "low cholesterol," and "low fat"[c]	21 CFR 101.82
Stanols/sterols and risk of coronary heart disease	Diets low in saturated fat and cholesterol that include two servings of foods that provide a daily total of at least 1.3 g of vegetable oil sterol esters in two meals may reduce the risk of heart disease. A serving of [name of the food] supplies __ g of vegetable oil sterol esters.	Must contain at least 0.65 g of β-sitosterol, campesterol, and stigmasterol (combined) or 1.7 g sitostanol and campestanol (combined) per reference amount and meet requirements for "low saturated fat," "low cholesterol," and "low fat" (unless qualified in dressings and spreads)	21 CFR 101.83

[a] "Good source" = contains at least 20% of the daily value of a particular nutrient per serving or reference amount; "low fat" = <3 g fat per reference amount customarily consumed (RACC) or 50 g if the serving size is small or <3 g fat per 100 g and <30% of calories from fat for meals; "low saturated fat" = <1 g saturated fat per RACC or 50 g if the serving size is small or <1 g saturated fat and <10% of calories from saturated fat for meals; "low cholesterol" = <20 mg cholesterol per RACC or 50 g if serving size is small or <20 mg cholesterol per 100 g for meals.

[b] CFR = Code of Federal Regulations.

[c] Requirements for low fat may be waived if the fat content exceeds the requirement due to fat naturally occurring in the ingredient added to make the health claim.

purified forms of β-glucan that have been extracted from flour. Regardless, a number of researchers have reported the effects of β-glucan addition to bread dough.[32,134–137] Skendi et al.[137] added a highly purified β-glucan of two molecular weights (1.0×10^5 and 2.03×10^5) to bread dough at levels up to 1.4% (flour weight) to determine the specific effects of β-glucan molecular weight on dough and bread properties. β-Glucan tended to increase water absorption and development time as fortification level increased. In the final product, β-glucan increased loaf volume up to 0.6–1.0% and decreased thereafter. It also produced a softer crumb, darker crumb, and more variation in gas cell size.

While there may be exceptions under some circumstances, in general the addition of β-glucan to bread dough, particularly when added in impure forms, has a detrimental effect on bread quality. Nevertheless, those interested in specialty products with specific health claims may be willing to sacrifice some sensory attributes for the health benefits imparted by this and other ingredients. Indeed, the addition of a poorly soluble β-glucan preparation to bread significantly decreased loaf volume, loaf height, increased firmness, and produced darker bread; however, when consumers were provided with the health information related to β-glucan, sensory scores for the fortified breads met or exceeded scores obtained for the control bread in appearance, flavor, and overall acceptability.[116]

Other researchers have sought to increase the nutritional quality of bread through the addition of other types of dietary fiber. Resistant starch is one type of dietary fiber that has found commercial success. Most likely this is because substantial additions (10–20% flour weight) of resistant starch to bread can be made, resulting in 5–10% total dietary fiber in the finished product, without considerable changes in bread quality or sensory perception.[138–140]

Prebiotics may also be incorporated into bread. Prebiotics are dietary fibers that result in the selective stimulation of beneficial bacteria in the gut at the expense of detrimental bacteria and thus improve host heath.[141] One of the most established types of prebiotics, fructans, have been added to bread.[142,143] Fructans are available commercially as short-chain fructooligosaccharides (degree of polymerization (DP) < 10), medium-chain inulin (DP < 40), long-chain inulin (DP < 100), or mixtures of these three. Peressini and Sensidoni[143] added fructans of average DP 10 and 23 to bread at 2.5–7.5% (flour weight). Fructans with average DP 10 exhibited a greater effect on water absorption, while the higher DP fructans had a greater effect on the viscoelastic properties of the dough. In moderately strong flour, the average DP 23 inulin resulted in reduced loaf volume and increased crumb firmness, while the opposite was true for average DP 10 inulin. In weak flour, both inulin products resulted in detrimental effects on loaf volume and firmness.

Another prebiotic that may be an interesting addition to bread is arabinoxylan oligosaccharides. Rather than adding these oligosaccharides to bread, however, they may be produced *in situ* during bread production through the action of xylanases. As mentioned, xylanases are available as dough enhancers to selectively hydrolyze the insoluble arabinoxylan, which is naturally present in wheat flour and imparts negative effects on bread-making quality.[144] Using higher than normal amounts of these enzymes, particularly if the enzymes are heat stable, can hydrolyze the arabinoxylan during the initial stages of the baking process and produce bread with substantial quantities of arabinoxylan oligosaccharides.[145]

Other functional dietary fibers may be used in bread to attract certain consumers. For instance, Lin et al.[146] added shiitake stipe (stem) to bread. Although sensory scores indicated that shiitake stipe reduced acceptability, the authors pointed out that this may be an interesting way of incorporating chitin, an amidated, insoluble polysaccharide, into foods. Chitin is not currently a food additive in the United States; however, it is taken by some of the population as a supplement for health reasons. Some consumers may be interested in products that contain naturally occurring chitin.

Increasing the antioxidant activity of baked goods has been a focus of much research. This is usually accomplished by adding an ingredient that is high in antioxidants to the formulation, which usually has undesirable effects on product quality or substantial effects on flavor (Table 4.5). Purified compounds may also be added. These often have less of an effect on the appearance and flavor of the product.

Other Baked Goods

As shown in Table 4.5, baked goods such as cookies, biscuits, crackers, and muffins have been supplemented with various ingredients to enhance nutrition. As with bread, many of these products have focused on improving gastrointestinal health or reducing inflammation. Because these products do not rely on gluten development for proper product quality, the overall sensory scores are often minimally impacted by the addition of nutraceuticals[127,128,130] or even enhanced.[109,132]

Summary

Breads and other baked goods are convenient and staple foods in the U.S. diet and as such offer an ideal venue as specialty foods. Because many specialty food consumers are on the lookout for healthy foods, we have discussed strategies to produce healthy baked goods in the whole wheat, multigrain, sourdough, organic, gluten-free, reduced sodium, and functional food categories. These categories represent major food trends in the area of baked goods. New strategies continue to develop in these areas that enhance product quality, nutrition, and sensory properties.

TABLE 4.5
Some Examples of Ingredients Added to Baked Goods to Produce Functional Foods and Their Effects on Product Quality

Product	Ingredient	Purpose	Amount	Effects on Product[a]	Reference
Biscuit	Red palm olein	Increase tocopherols, carotenoids, and antioxidant activity	40–60% replacement of shortening	↓ Water loss during baking, ↓ specific volume, ↑ darkness, ↓ water activity; ↓ shearing force, ↑ overall sensory scores	109
Bread	Elemental iron, ferric sodium ethylenediaminetetraacetate, zinc sulfate or zinc oxide	Improve mineral nutrition	Up to 60 mg/kg of flour	↑ Water absorption, ↓ dough development time	110
Bread	Green tea extract	Increase antioxidant activity	Up to 0.8% recommended	↓ Sensory aroma, ↓ flavor and overall acceptability at high levels; minimal effect at recommended level	111
Bread	Mushroom powder	Increase vitamin D, selenium, chromium, and antioxidants	Up to 20% (flour weight)	↓ Loaf volume, ↑ firmness	112
Bread	Green coffee extract	Improve chemoprotective effects	1%	↑ Antioxidant activity, ↑ resistance of colon and liver cells against hydrogen peroxide-mediated genotoxicity	113
Bread	Sweet potato	Increase carotenoid content	Sweet potato added to reach 15 μg β-carotene/g bread (good source)	Produced a yellowish, heavier bread that was preferred by consumers	114
Bread	Freeze dried banana flour	Increase potassium content	Up to 30% flour replacement (128.6 mg of potassium/30 g serving)	No change in loaf volume when flour substituted with 25% gluten; ↑ darkness; ↑ water absorption; ↓ mixing tolerance; ↑ staling rate	115

Food	Compound	Purpose	Amount	Effect	Ref.
Bread	β-Glucan	Increase soluble fiber; make health claim regarding coronary heart disease (Table 4.4)	Up to 1.5 g β-glucan/serving of bread	↓ Loaf volume, ↑ firmness, ↑ darkness	116
Bread	Vitamin D (cholecalciferol)	Improve vitamin D status of women	3.6 μg/30 g serving	Vitamin D was stable during bread making and improved vitamin D status	117
Bread	Grape seed extract	Increase anthocyanin content and antioxidant activity	Up to 0.2% of flour weight	Minor darkening of color; no effects on bread quality; up to 40% increase in antioxidant activity	118
Bread	Soy flour and soy protein isolate	Increase soy isoflavone content	Up to 5.5 μmol total isoflavones/g dry bread	Soy isoflavones were stable during bread making	119
Bread	Kiwi fruit extract	Increase antioxidant activity and vitamin C content	About 15% in final product	↑ Fruit flavor and aroma of bread; ↑ wetness; ↓ hardness and coarseness; no change in overall impression of breads	120
Bread	Calcium sulfate, calcium carbonate, calcium citrate	Increase calcium content	1.25 g calcium carbonate, 1.77 g calcium sulfate, or 1.16 g calcium citrate/100 g flour	Recommended fortification levels are not significantly different from regular bread (triangle test) and provide 150, 115, or 75 mg calcium per 30 g portion for calcium carbonate, calcium sulfate, and calcium citrate, respectively	121
Bread, cookies	Turmeric powder	Increase antioxidant and anti-inflammatory properties	Up to 4% in bread; 6% in cookies	↑ density; ↑ hardness, ↑ darkness, no significant difference between 6% addition and control cookies	122
Bread, cookies, muffins	Corn flour/high lutein wheat/ free lutein	Increase lutein content to promote eye health	Up to 1 mg/serving recommended to reach 5–6 mg/d (amount required for eye health)	Corn and high lutein wheat flour caused darkening; substantial lutein loss during baking; lutein more stable in bread than cookies and muffins	123

(Continued)

TABLE 4.5 (CONTINUED)
Some Examples of Ingredients Added to Baked Goods to Produce Functional Foods and Their Effects on Product Quality

Product	Ingredient	Purpose	Amount	Effects on Product[a]	Reference
Cake	Turmeric powder	Increase antioxidant and anti-inflammatory properties	6% in final formulation recommended	Sensory scores comparable when turmeric powder was used at recommended concentration	124
Cake	Turmeric powder	Increase antioxidant and anti-inflammatory properties	1.6% in final formulation recommended	Sensory scores comparable when turmeric powder was used at recommended concentration	125
Cookies	Cumin and ginger	Increase total phenolic content and antioxidant activity	5%	↓ Dough stability, ↑ spread ratios, ↑ softness, ↑ darkness	126
Cookies	ROPUFA 10 (DSM Inc.).	Increase long-chain ω-3 fatty acids (eicosapentaenoic and docosahexaenoic acids)	400 mg long-chain ω-3 fatty acids/cookie	No losses in ω-3 fatty acids during 28 d of storage; no significant differences from control cookies	127
Cookies	Isochrysis galbana (algae) biomass	Increase long-chain ω-3 fatty acids (eicosapentaenoic and docosahexaenoic acids)	Up to 96 mg long-chain ω-3 fatty acids/30 g cookie	Cookie quality was maintained in cookies	128
Cookies	Fructooligosaccharides	Prebiotic source, reduce calories	20–60% replacement of sugar	Most acceptable product with 60% fructooligosaccharides replacing sugar; 5.2 g dietary fiber/30 g serving and 11% fewer calories	40

Cookies	Blue corn flour	Increase anthocyanin content due to health benefits	80% replacement of wheat flour	Adding an acid and using shorter baking times minimized anthocyanin degradation during baking but caused the cookies to turn pink	129
Muffins	Phytosterols (unidentified source)	Heart disease prevention	0.94 g total phytosterols/30 g serving	No significant difference from control muffins (triangle test)	130
Muffins	Oat bran (β-glucan)	Reduce postprandial blood glucose	2 or 4 g β-glucan/muffin	High molecular weight β-glucan reduced postprandial blood glucose area under the curve by 44% compared to control whole wheat muffin; low molecular weight only decreased blood glucose by 15%	131
Muffins	Commercial inulin powder and gel	Replace fat with inulin, a prebiotic	Up to 20% of flour weight	↑ Moisture, ↑ density, ↓ volume, ↑ sensory toughness, ↓ sensory flavor	132

a ↑/↓ = increased/decreased compared with control without added functional ingredient.

References

1. National Association for Specialty Food Trade. 2010. Liven up lunch boxes with distinctive snacks. National Association for the Specialty Food Trade offers picks for school. http://www.virtualpressoffice.com/detail.do?contentId=351506&showId=1206207880303 (accessed May 9, 2011).
2. Sloan, A.E. 2010. Giving consumers what they want. *Food Technology* 64: 52–53.
3. U.S. Department of Agriculture. 2010. *Dietary guidelines for Americans, 2010.* http://www.cnpp.usda.gov/Publications/DietaryGuidelines/2010/PolicyDoc/PolicyDoc.pdf (accessed May 9, 2011).
4. Vocke, G., Buzby, J.C., and Wells, H.F. 2008. Consumer preferences change wheat flour use. *Amber Waves* 6: 2.
5. Gan, Z., Galliard, T., Ellis, P.R., Angold, R.E., and Vaughan, J.G. 1992. Effect of the outer bran layers on the loaf volume of wheat bread. *Journal of Cereal Science* 15: 151–163.
6. Zhang, D., and Moore, W.R. 1999. Wheat bran particle size effects on bread baking performance and quality. *Journal of the Science of Food and Agriculture* 79: 805–809.
7. Bakke, A., and Vickers, Z. 2007. Consumer liking of refined and whole wheat breads. *Journal of Food Science* 72: S473–S480.
8. Pawar, V.D., and Machewad, G.M. 2004. Organoleptic quality of bread enriched with high protein products from sunflower meal. *Journal of Food Science and Technology—Mysore* 41: 672–674.
9. Rawat, A., Singh, G., Mital, B.K., and Mittal, S.K. 1994. Effect of soy-fortification on quality characteristics of chapatis. *Journal of Food Science and Technology—Mysore* 31: 114–116.
10. Shogren, R.L., Mohamed, A.A., and Carriere, C.J. 2003. Sensory analysis of whole wheat/soy flour breads. *Journal of Food Science* 68: 2141–2145.
11. Busken, D.F. 2011. Understanding the challenges of gluten-free baking. *Cereal Foods World* 56: 42–43.
12. Stampfli, L., and Nersten, B. 1995. Emulsifiers in breadmaking. *Food Chemistry* 52: 353–360.
13. Andersson, A.A.M., Landberg, R., Soderman, T. et al. 2011. Effects of alkylres-orcinols on volume and structure of yeast-leavened bread. *Journal of the Science of Food and Agriculture* 91: 226–232.
14. Boz, H., Karaoglu, M.M., Kotancilar, H.G., and Gercekaslan, K.E. 2010. The effects of different materials as dough improvers for organic whole wheat bread. *International Journal of Food Science and Technology* 45: 1472–1477.
15. Grausbruber, H., Meisenberger, S., Schoenlechner, R., and Vollmann, J. 2008. Influence of dough improvers on whole-grain bread quality of einkorn wheat. *Acta Alimentaria* 37: 379–390.
16. Gerrard, J.A., and Sutton, K.H. 2005. Addition of transglutaminase to cereal products may generate the epitope responsible for coeliac disease. *Trends in Food Science and Technology* 16: 510–512.
17. Dekking, E.H.A., Van Veelen, P.A. de Ru, A. et al. 2008. Microbial transgluta-minases generate T cell stimulatory epitopes involved in celiac disease. *Journal of Cereal Science* 47: 339–346.

18. Seguchi, M., Uozu, M., Oneda, H., Murayama, R., and Okusu, H. 2010. Effect of outer bran layers from germinated wheat grains on breadmaking properties. *Cereal Chemistry* 87: 231–236.
19. McGuire, C.F., and Opalka, J. 1995. Sensory evaluation of a hard white compared to a hard red winter wheat. *Journal of the Science of Food and Agriculture* 67: 129–133.
20. Kihlberg, I., Johansson, L., Kohler, A., and Risvik, E. 2004. Sensory qualities of whole wheat pan bread—Influence of farming system, milling and baking technique. *Journal of Cereal Science* 39: 67–84.
21. de Kock, S., Taylor, J., and Taylor, J.R.N. 1999. Effect of heat treatment and particle size of different brans on loaf volume of brown bread. *Lebensmittel Wissenschaft und Technologie* 32: 349–356.
22. Noort, M.W.J., van Haaster, D., Hemery, Y., Schols, H.A., and Hamer, R.J. 2010. The effect of particle size of wheat bran fractions on bread quality—Evidence for fibre protein interactions. *Journal of Cereal Science* 52: 59–64.
23. Camire, M.E., Bolton, J., Jordan, J.J. et al. 2006. Color influences consumer opinions of wheat muffins. *Cereal Foods World* 51: 274–276.
24. Gaines, C.S., and Donelson, J.R. 1985. Evaluating cookie spread potential of whole wheat flours from soft wheat cultivars. *Cereal Chemistry* 62: 134–136.
25. Gomez, M., Manchon, L., Oliete, B., Ruiz, E., and Caballero, P.A. 2010. Adequacy of wholegrain non-wheat flours for layer cake elaboration. *LWT— Food Science and Technology* 43: 507–513.
26. Masoodi, F.A., Chauhan, G.S., Tyagi, S.M., Kumbhar, B.K., and Kaur, H. 2001. Effect of apple pomace incorporation on rheological characteristics of wheat flour. *International Journal of Food Properties* 4: 215–223.
27. Hu, G., Huang, S., Cao, S., and Ma, Z. 2009. Effect of enrichment with hemicellulose from rice bran on chemical and functional properties of bread. *Food Chemistry* 115: 839–842.
28. Bouaziz, M.A., Amara, W.B., Attia, H., Blecker, C., and Besbes, S. 2010. Effect of the addition of defatted date seeds on wheat dough performance and bread quality. *Journal of Texture Studies* 41: 511–531.
29. Indrani, D., Rajiv, J., and Rao, G.V. 2010. Influence of fenugreek seed powder on the dough rheology, microstructure and quality of parotta—An Indian flat bread. *Journal of Texture Studies* 41: 208–223.
30. Lin, L.Y., Hsieh, Y.J., Liu, H.M., Lee, C.C., and Mau, J.L. 2009. Flavor components in buckwheat bread. *Journal of Food Processing and Preservation* 33: 814–826.
31. Bodroza-Solarov, M., Filipcev, B., Kevresan, Z., Mandic, A., and Simurina, O. 2008. Quality of bread supplemented with popped *Amaranthus cruentus* grain. *Journal of Food Process Engineering* 31: 602–618.
32. Sullivan, P., O'Flaherty, J., Brunton, N., Arendt, E., and Gallagher, E. 2010. Fundamental rheological and textural properties of doughs and breads produced from milled pearled barley flour. *European Food Research and Technology* 231: 441–453.
33. Pohjanheimo, T.A., Hakala, M.A., Tahvonen, R.L., Salminen, S.J., and Kallio, H.P. 2006. Flaxseed in breadmaking: Effects on sensory quality, aging, and composition of bakery products. *Journal of Food Science* 71: S343–S348.
34. Losso, J.N., Holliday, D.L., Finley, J.W., et al. 2009. Fenugreek bread: A treatment for diabetes mellitus. *Journal of Medicinal Food* 12: 1046–1049.

35. Sharma, H.R., and Chauhan, G.S. 2002. Effects of stabilized rice bran—Fenugreek blends on the quality of breads and cookies. *Journal of Food Science and Technology—Mysore* 39: 225–233.
36. Skrbic, B., and Cvejanov, J. 2011. The enrichment of wheat cookies with high-oleic sunflower seed and hull-less barley flour: Impact on nutritional composition, content of heavy elements and physical properties. *Food Chemistry* 124: 1416–1422.
37. Filipcev, B., Simurina, O., Sakac, M. et al. 2011. Feasibility of use of buckwheat flour as an ingredient in ginger nut biscuit formulation. *Food Chemistry* 125: 164–170.
38. Hooda, S., and Jood, S. 2005. Organoleptic and nutritional evaluation of wheat biscuits supplemented with untreated and treated fenugreek flour. *Food Chemistry* 90: 427–435.
39. Moraes, E.A., Dantas, M.I.D., Morais, D.D. et al. 2010. Sensory evaluation and nutritional value of cakes prepared with whole flaxseed flour. *Ciencia e Tecnologia de Alimentos* 30: 974–979.
40. Handa, C., Goomer, S., Siddhu, A. 2011. Effects of whole-multigrain and fructoligosaccharide incorporation on the quality and sensory attributes of cookies. *Food Science and Technology Research* 17: 45–54.
41. Salim-ur-Rehman, Paterson, A., and Piggott, J.R. 2006. Flavour in sourdough breads: A review. *Trends in Food Science and Technology* 17: 557–566.
42. Brummer, J.M., and Lorenz, K. 2003. Preferments and sourdoughs in German breads. In *Handbook of dough fermentations*, ed. K. Kulp and K. Lorenz, 247–267. New York: Marcel Dekker.
43. Clarke, C.I., and Arendt, E. 2005. A review of the application of sourdough technology to wheat breads. *Advances in Food and Nutrition Research* 49: 137–161.
44. Arendt, E., Ryan, A., and Dal Bello, F. 2007. Impact of sourdough on the texture of bread. *Food Microbiology* 24: 165–174.
45. Poutanen, K., Flander, L., and Katina, K. 2009. Sourdough and cereal fermentation in a nutritional perspective. *Food Microbiology* 26: 693–699.
46. Salmenkallio-Marttila, M., Katina, K., and Autio, K. 2001. Effect of bran fermentation on quality and microstructure of high-fibre wheat bread. *Cereal Chemistry* 78: 429–435.
47. Gobetti, M., Corsetti, A., and Rossi, J. 1995. Interaction between lactic acid bacteria and yeasts in sourdough using a rheofermentometer. *World Journal of Microbiology and Biotechnology* 11: 625–630.
48. Foster-Powell, K., Holt, S.H.A., and Brand-Miller, J.C. 2002. International table of glycemic index and glycemic load values: 2002. *American Journal of Clinical Nutrition* 76: 5–56.
49. Scazzina, F., Del Rio, D., Pellegrini, N., and Brighenti, F. 2009. Sourdough bread: Starch digestibility and postprandial glycemic response. *Journal of Cereal Science* 49: 419–421.
50. Lappi, J., Selinheimo, E., Schwab, U. et al. 2010. Sourdough fermentation of wholemeal wheat bread increases solubility of arabinoxylan and protein and decreases postprandial glucose and insulin responses. *Journal of Cereal Science* 51: 152–158.
51. Lopez, H.W., Leenhardt, F., and Remesy, C. 2004. New data on the bioavailability of bread magnesium. *Magnesium Research* 17: 335–340.
52. Leenhardt, F., Levrat-Verny, M.A., Chanliaud, E., and Remesy, C. 2002. Moderate decrease of pH by sourdough fermentation is sufficient to reduce

phytate content of whole wheat flour through endogenous phytase activity. *Journal of Agricultural and Food Chemistry* 53: 98–102.

53. Martinez-Anaya, M.A., Devesa, A., Andreau, P., Escriva, C., and Collar, C. 1998. Effects of the combination of starters and enzymes in regulating bread quality and shelf life. *Food Science and Technology International* 4: 425–435.

54. Fazeli, M.R., Shahverdi, A.R., Sedaghat, B., Jamalifar, H., and Samadi, N. 2004. Sourdough-isolated *Lactobacillus fermentum* as a potent anti-mould preservative of a traditional Iranian bread. *European Food Research and Technology* 218: 554–556.

55. Dal Bello, F., Clarke, C.I., Ryan, L.A.M. et al. 2007. Improvement of the quality and shelf life of wheat bread by fermentation with the antifungal strain *Lactobacillus plantarum* FST 1.7. *Journal of Cereal Science* 45: 309–318.

56. Zhang, C.G., Brandt, M.J., Schwab, C., and Ganzle, M.G. 2010. Propionic acid production by cofermentation of *Lactobacillus buchneri* and *Lactobacillus diolivorans* in sourdough. *Food Microbiology* 27: 390–395.

57. Rogers, D.E., and Hoseney, R.C. 1989. Effects of fermentation in saltine cracker production. *Cereal Chemistry* 66: 1–10.

58. Katina, K., Liukkonen, K.H., Kaukovirta-Norja, A., et al. 2007. Fermentation-induced changes in the nutritional value of native or germinated rye. *Journal of Cereal Science* 46: 348–355.

59. Gobbetti, M., Di Cagno, R., and De Angelis, M. 2010. Functional microorganisms for functional food quality. *Critical Reviews in Food Science and Nutrition* 50: 716–727.

60. Certel, M., Erbas, M., Uslu, M.K., and Erbas, M.O. 2007. Effects of fermentation time and storage on the water-soluble vitamin contents of tarhana. *Journal of the Science of Food and Agriculture* 87: 1215–1218.

61. L-Baeckstrom, G., Hanell, U., and Svensson, G. 2004. Baking quality of winter wheat grown in different cultivating systems, 1992–2001: A holistic approach. *Journal of Sustainable Agriculture* 24: 53–79.

62. Carcea, M., Salvatorelli, S., Turfani, V., and Mellara, F. 2006. Influence of growing conditions on the technological performance of bread wheat (*Triticum aestivum* L.). *International Journal of Food Science and Technology* 41: 102–107.

63. Annett, L.E., Spaner, D., and Wismer, W.V. 2007. Sensory profiles of bread made from paired samples of organic and conventionally grown wheat grain. *Journal of Food Science* 72: S254–S260.

64. Gelinas, P., Morin, C., Reid, J.F., and Lachance, P. 2009. Wheat cultivars grown under organic agriculture and the bread making performance of stone-ground whole wheat flour. *International Journal of Food Science and Technology* 44: 525–530.

65. Krejcirova, L., Capouchova, I., Petr, J., Bicanova, E., and Famera, O. 2007. The effect of organic and conventional growing systems on quality and storage protein composition of winter wheat. *Plant Soil and Environment* 53: 499–505.

66. Nitika, D.P., and Khetarpaul, N. 2008. Physico-chemical characteristics, nutrient composition and consumer acceptability of wheat varieties grown under organic and inorganic farming conditions. *International Journal of Food Sciences and Nutrition* 59: 224–245.

67. Catassi, C., and Yachha, S.K. 2006. The epidemiology of celiac disease. In *The science of gluten-free foods and beverages*, ed. E.K. Arendt and F. Dal Bello, 1–13. St. Paul, MN: AACC International.

68. Rubio-Tapia, A., Kyle, R.A., Kaplan, E.L. et al. 2009. Increased prevalence and mortality in undiagnosed celiac disease. *Gastroenterology* 137: 88–93.
69. Packaged Facts. The gluten-free food and beverage market: Trends and developments worldwide. 2nd ed. Rockville, MD: Packaged Facts. http://www.specialtyfood.com/news-trends/featured-articles/retail-operations/gotta-have-gluten-free/.
70. Hartman Group. 2009. Making sense of the gluten-free trend. http://www.hartman-group.com/hartbeat/making-sense-of-the-gluten-free-trend (accessed November 22, 2010).
71. Mariani, P., Grazia, V.M., Montouri, M. et al. 1998. The gluten-free diet: A nutritional risk factor for adolescents with celiac disease. *Journal of Pediatric Gastroenterology and Nutrition* 27: 519–523.
72. Gallagher, E., Gormley, T.R., and Arendt, E.K. 2004. Recent advances in the formulation of gluten-free cereal-based products. *Trends in Food Science and Technology* 15: 143–152.
73. Ahlborn, G.J., Pike, O.A., Hendrix, S.B., Hess, W.M., and Huber, C.S. 2005. Sensory, mechanical, and microscopic evaluation of staling in low-protein and gluten-free breads. *Cereal Chemistry* 82: 328–335.
74. Gambus, H., Nowotna, A., Ziobro, R., Gumul, D., and Sikora, M. 2001. The effect of use of guar gum with pectin mixture in gluten-free bread. *Electronic Journal of Polish Agricultural Universities* 4: 1–13.
75. Kadan, R.S., Robinson, M.G., Thibodeaux, D.P., and Pepperman, A.B. 2001. Texture and other physicochemical properties of whole rice bread. *Journal of Food Science* 66: 940–944.
76. Lazaridou, A., Duta, D., Papageorgiou, M., Belc, N., and Biliaderis, C.G. 2007. Effects of hydrocolloids on dough rheology and bread quality parameters in gluten-free formulations. *Journal of Food Engineering* 79: 1033–1047.
77. McCarthy, D.F., Gallagher, E., Gormley, T.R., Schober, T.J., and Arendt, E.K. 2005. Application of response surface methodology in the development of gluten-free bread. *Cereal Chemistry* 82: 609–615.
78. Moore, M.M., Schober, T.J., Dockery, P., and Arendt, E.K. 2004. Textural comparisons of gluten-free and wheat-based doughs, batters, and breads. *Cereal Chemistry* 81: 567–575.
79. Sabanis, D., Lebesi, D., and Tzia, C. 2009. Development of fibre-enriched gluten-free bread: A response surface methodology study. *International Journal of Food Sciences and Nutrition* 60: 174–190.
80. Schober, T.J., Bean, S.R., Boyle, D.L., and Park, S.H. 2008. Improved viscoelastic zein–starch doughs for leavened gluten-free breads: Their rheology and microstructure. *Journal of Cereal Science* 48: 755–767.
81. Sivaramakrishnan, H.P., Senge, B., and Chattopadhyay, P.K. 2004. Rheological properties of rice dough for making rice bread. *Journal of Food Engineering* 62: 37–45.
82. Toufeili, I., Dagher, S., Shadarevian, S., Noureddinei, A., Sarakbi, M., and Ferran, M.T. 1994. Formulation of gluten-free pocket-type flat breads: Optimization of methylcellulose, gum arabic, and egg albumen levels by response surface methodology. *Cereal Chemistry* 71: 594–601.
83. Peressini, D., Pin, M., and Sensidoni, A. 2011. Rheology and breadmaking performance of rice-buckwheat batters supplemented with hydrocolloids. *Food Hydrocolloids* 25: 340–349.

84. Bell, D.A. 1990. Methylcellulose as a structure enhancer in bread baking. *Cereal Foods World* 35: 1001–1006.

85. Kobylanski, J.R., Perez, O.E., and Pilosof, A.M.R. 2004. Thermal transitions of gluten-free doughs as affected by water, egg white and hydroxypropylmethylcellulose. *Thermochim Acta* 411: 81–89.

86. Alvarez-Jubete, L., Auty, M., Arendt, E.K., and Gallagher, E. 2010a. Baking properties and microstructure of pseudocereal flours in gluten-free bread formulations. *European Food Research and Technology* 230: 437–445.

87. Alvarez-Jubete, L., Arendt, E.K., and Gallagher, E. 2010b. Nutritive value of pseudocereals and their increasing use as functional gluten-free ingredients. *Trends in Food Science and Technology* 21: 106–113.

88. Mariotti, M., Lucisano, M., Pagani, M.A., and Ng, P.K.W. 2009. The role of corn starch, amaranth flour, pea isolate, and psyllium flour on the rheological properties and the ultrastructure of gluten-free doughs. *Food Research International* 42: 963–975.

89. De la Barca, A.M.C., Rojas-Martinez, M.E., Islas-Rubio, A.R., and Cabrera-Chavez, F. 2010. Gluten-free breads and cookies of raw and popped amaranth flours with attractive technological and nutritional qualities. *Plant Foods for Human Nutrition* 65: 241–246.

90. Huttner, E.K., and Arendt, E.K. 2010a. Recent advances in gluten-free baking and the current status of oats. *Trends in Food Science and Technology* 21: 303–312.

91. Huttner, E.K., Del Bello, F., and Arendt, E.K. 2010b. Rheological properties and breadmaking performance of commercial wholegrain oat flours. *Journal of Cereal Science* 52: 65–71.

92. Moroni, A.V., Dal Bello, F., and Arendt, E.K. 2009. Sourdough in gluten-free bread-making: An ancient technology to solve a novel issue? *Food Microbiology* 26: 676–684.

93. Gujral, H.S., and Rosell, C.M. 2004. Functionality of rice flour modified with a microbial transglutaminase. *Journal of Cereal Science* 39: 225–230.

94. Renzetti, S., and Arendt, E.K. 2009. Effects of oxidase and protease treatments on the breadmaking functionality of a range of gluten-free flours. *European Food Research and Technology* 229: 307–317.

95. Ronda, F., Gomez, M., Caballero, P.A., Oliete, B., and Blanco, C.A. 2009. Improvement of quality of gluten-free layer cakes. *Food Science and Technology International* 15: 193–202.

96. Sumnu, G., Koksel, F., Sahin, S., Basman, A., and Meda, V. 2010. The effects of xanthan and guar gums on staling of gluten-free rice cakes baked in different ovens. *International Journal of Food Science and Technology* 45: 87–93.

97. Gambus, H., Gambus, F., Pastuszka, D. et al. 2009. Quality of gluten-free supplemented cakes and biscuits. *International Journal of Food Sciences and Nutrition* 60: 31–50.

98. Sedej, I., Saka, M., Mandi, A. et al. 2011. Quality assessment of gluten-free crackers based on buckwheat flour. *LWT—Food Science and Technology* 44: 694–699.

99. Han, J., Janz, J.A.M., and Gerlat, M. 2010. Development of gluten-free cracker snacks using pulse flours and fractions. *Food Research International* 43: 627–633.

100. Food and Nutrition Board. 2005. Sodium and chloride. In *Dietary Reference Intakes for water, potassium, sodium, chloride, and sulfate*, ed. Institute of Medicine, 269–243. Washington DC: National Academies Press.
101. International Food Information Council. 2009. Consumer sodium research: Concern, perceptions and action. http://www.foodinsight.org/Content/6/FINAL-IFIC-Sodium-Consumer-Research-Report-8-14-09.pdf (accessed Feburary 8, 2011).
102. National Cancer Institute. 2010. Sources of sodium in the diets of the U.S. population ages 2 years and older, NHANES 2005–2006. *Risk Factor monitoring and methods, cancer control and population sciences*. http://riskfactor.cancer.gov/diet/foodsources/sodium/table1a.html (accessed February 8, 2011).
103. Heidolph, B.B., Ray, D.K., Roller, S. et al. 2011. Looking for my lost shaker of salt … replacer: Flavor, function, future. *Cereal Foods World* 56: 5–19.
104. Noort, M.W.J., Bult, J.H.F., Stieger, M., and Hamer, R.J. 2010. Saltiness enhancement in bread by inhomogeneous spatial distribution of sodium chloride. *Journal of Cereal Science* 52: 378–386.
105. Salovaara, H. 1982. Sensory limitations to replacement of sodium with potassium and magnesium in bread. *Cereal Chemistry* 59: 427–430.
106. Charlton, K.E., MacGregor, E., Vorster, N.H., Levitt, N.S., and Steyn, K. 2007. Partial replacement of NaCl can be achieved with potassium, magnesium and calcium salts in brown bread. *International Journal of Food Sciences and Nutrition* 58: 508–521.
107. Braschi, A., Gill, L., and Naismith, D.J. 2009. Partial substitution of sodium with potassium in white bread: Feasibility and bioavailability. *International Journal of Food Sciences and Nutrition* 60: 507–521.
108. Doyon, M., and Labrecque, J. 2008. Functional foods: A conceptual definition. *British Food Journal* 110: 1133–1149.
109. El-Hadad, N., Abou-Gharbia, H.A., El-Aal, M.H.A., and Youssef, M.M. 2010. Red palm olein: Characterization and utilization in formulating novel functional biscuits. *Journal of the American Oil Chemists' Society* 87: 295–304.
110. Akhtar, S., Anjum, F., Saleem-Ur-Rehman, and Sheikh, M.A. 2009. Effect of mineral fortification on rheological properties of whole wheat flour. *Journal of Texture Studies* 40: 51–65.
111. Bajerska, J., Mildner-Szkudlarz, S., Jeszka, J., and Szwengiel, A. 2010. Catechin stability, antioxidant properties and sensory profiles of rye breads fortified with green tea extracts. *Journal of Food and Nutrition Research* 49: 104–111.
112. Corey, M.E., Beelman, R.B., and Seetharman, K. 2009. Potential for nutritional enrichment of whole-wheat bread with portabella mushroom powder (*Agaricus bisporus* (J. Lge) Imbach, Agaricomycetideae). *International Journal of Medicinal Mushrooms* 11: 157–166.
113. Glei, M., Kirmse, A., Habermann, N., Persin, C., and Pool-Zobel, B.L. 2006. Bread enriched with green coffee extract has chemoprotective and antigenotoxic activities in human cells. *Nutrition and Cancer* 56: 182–192.
114. Low, J.W., and van Jaarsveld, P.J. The potential contribution of bread buns fortified with beta-carotene-rich sweet potato in Central Mozambique. *Food and Nutrition Bulletin* 29: 98–107.
115. Mohamed, A., Xu, J., and Singh, M. 2010. Yeast leavened banana-bread: Formulation, processing, colour and texture analysis. *Food Chemistry* 118: 620–626.

116. Moriartey, S., Temelli, F., Vasanthan, T. 2010. Effect of health information on consumer acceptability of bread fortified with beta-glucan and effect of fortification on bread quality. *Cereal Chemistry* 87: 428–433.

117. Natri, A.M., Salo, P., Vikstedt, T. et al. 2006. Bread fortified with cholecalciferol increases the serum 25-hydroxyvitamin D concentration in women as effectively as a cholcalciferol supplement. *Journal of Nutrition* 136: 123–127.

118. Peng, X., Maa, J., Cheng, K.W., Jiang, Y., Chen, F., and Wang, M. 2010. The effects of grape seed extract fortification on the antioxidant activity and quality attributes of bread. *Food Chemistry* 119: 49–53.

119. Shao, S.Q., Duncan, A.M., Yang, R., Marcone, M.F., Rajcan, I., and Tsao, R. 2009. Tracking isoflavones: From soybean to soy flour, soy protein isolates to functional soy bread. *Journal of Functional Foods* 1: 119–127.

120. Sun-Waterhouse, D., Chen, J., Chuah, C. et al. 2009. Kiwifruit-based polyphenols and related antioxidants for functional foods: Kiwifruit extract-enhanced gluten-free bread. *International Journal of Food Sciences and Nutrition* 60: 251–264.

121. Ziadeh, G., Shadarevian, S., Malek, A. et al. 2005. Determination of sensory thresholds of selected calcium salts and formulation of calcium-fortified pocket-type flat bread. *Journal of Food Science* 70: S548–S552.

122. Lim, H.S., Park, S.H., Ghafoor, K., Hwang, S.Y., and Park, J. 2011. Quality and antioxidant properties of bread containing turmeric (*Curcuma longa* L.) cultivated in South Korea. *Food Chemistry* 124: 1577–1582.

123. Abdel-Aal, E.S.M., Young, J.C., Akhtar, H., and Rabalski, I. 2010. Stability of lutein in wholegrain bakery products naturally high in lutein or fortified with free lutein. *Journal of Agricultural and Food Chemistry* 58: 10109–10117.

124. Lim, H.S., Ghafoor, K., Park, S.H., Hwang, S.Y., and Park, J. 2010. Quality and antioxidant properties of yellow layer cake containing Korean turmeric (*Curcuma longa* L.) powder. *Journal of Food and Nutrition Research* 49: 123–133.

125. Seo, M.J., Park, J.E., and Jang, M.S. 2010. Optimization of sponge cake added with turmeric (*Curcuma longa* L.) powder using mixture design. *Food Science and Biotechnology* 19: 617–625.

126. Abdel-Samie, M.A.S., Wan, J.J., Huang, W.N., Chung, O.K., and Xu, B.C. 2010. Effects of cumin and ginger as antioxidants on dough mixing properties and cookie quality. *Cereal Chemistry* 87: 454–460.

127. Boreno, R., Kocer, D., Ghai, G., Tepper, B.J., and Karwe, M.V. 2007. Stability and consumer acceptance of long-chain omega-3 fatty acids (eicosapentaenoic acid, 20:5, n-3 and docosahexaenoic acid, 22:6, n-3) in cream-filled sandwich cookies. *Journal of Food Science* 72: S49–S54.

128. Gouveia, L., Coutinho, C., Mendonca, E. et al. 2008. Functional biscuits with PUFA-omega 3 from *Isochrysis galbana*. *Journal of the Science of Food and Agriculture* 88: 891–896.

129. Li, J., Walker, C.E., and Faubion, J.M. 2011. Acidulant and oven type affect total anthocyanin content of blue corn cookies. *Journal of the Science of Food and Agriculture* 91: 38–43.

130. Quilez, J., Ruiz, J.A., Brufau, G., and Rafecas, M. 2006. Bakery products enriched with phytosterols, α-tocopherol and β-carotene. Sensory evaluation and chemical comparison with market products. *Food Chemistry* 94: 399–405.

131. Tosh, S.M., Brummer, Y., Wolever, T.M.S., and Wood, P.J. 2008. Glycemic response to oat bran muffins treated to vary molecular weight of beta-glucan. *Cereal Chemistry* 85: 211–217.

132. Zhan, S., Pepke, F., and Rohm, H. 2010. Effect of inulin as a fat replacer on texture and sensory properties of muffins. *International Journal of Food Science and Technology* 45: 2531–2537.

133. Tiwari, U., and Cummins, E. 2009. Factors influencing β-glucan levels and molecular weight in cereal-based products. *Cereal Chemistry* 86: 290–301.

134. Ostman, E., Rossi, E., Larsson, H., Brighenti, F., and Bjorck, I. 2006. Glucose and insulin responses in healthy men to barley bread with different levels of (1 → 3; 1 → 4)-beta-glucans; predictions using fluidity measurements of *in vitro* enzyme digests. *Journal of Cereal Science* 43: 230–235.

135. Brennan, C.S., and Cleary, L.J. 2007. Utilisation Glucagel® in the β-glucan enrichment of breads: A physicochemical and nutritional evaluation. *Food Research International* 40: 291–296.

136. Cleary, L.J., Andersson, R., and Brennan, C.S. 2007. The behaviour and susceptibility to degradation of high and low molecular weight barley beta-glucan in wheat bread during baking and *in vitro* digestion. *Food Chemistry* 102: 889–897.

137. Skendi, A., Biliaderis, C.G., Papageorgiou, M., and Izydorczyk, M.S. 2010. Effects of two barley b-glucan isolates on wheat flour dough and bread properties. *Food Chemistry* 119: 1159–1167.

138. Van Hung, P., Yamamori, M., and Morita, N. 2005. Formation of enzyme-resistant starch in bread as affected by high-amylose wheat flour substitutions. *Cereal Chemistry* 82: 690–694.

139. Ozturk, S., Koksel, H., and Ng, P.K.W. 2009. Farinograph properties and bread quality of flours supplemented with resistant starch. *International Journal of Food Science and Nutrition* 60: 449–457.

140. Sanz-Penella, J.M., Wronkowska, M., Soral-Smietana, M., Collar, C., and Haros, M. Impact of the addition of resistant starch from modified pea starch on dough and bread performance. *European Food Research and Technology* 231: 499–508.

141. Gibson, G.R., and Roberfroid, M.B. 1995. Dietary modulation of the human colonic microbiota—Introducing the concept of prebiotics. *Journal of Nutrition* 125: 1401–1412.

142. Wang, J.S., Rosell, C.M., and de Barber, C.B. 2002. Effect of the addition of different fibres on wheat dough performance and bread quality. *Food Chemistry* 79: 221–226.

143. Peressini, D., and Sensidoni, A. 2009. Effect of soluble dietary fibre addition on rheological and breadmaking properties of wheat doughs. *Journal of Cereal Science* 49: 190–201.

144. Courtin, C.M., and Delcour, J.A. 2002. Arabinoxylans and endoxylanases in wheat flour bread-making. *Journal of Cereal Science* 35: 225–243.

145. Dornez, E., Verjans, P., Broekaert, W.F. et al. 2011. *In situ* production of prebiotic AXOS by hyperthermophilic xylanase B from *Thermotoga maritime* in high quality bread. *Cereal Chemistry* 88: 124–129.

146. Lin, L.Y., Tseng, Y.H., Li, R.C., and Mau, J.L. 2008. Quality of shiitake stipe bread. *Journal of Food Processing and Preservation* 32: 1002–1015.

5

Specialty Condiments, Dressings, and Sauces

Melissa Sales and Mark A. Daeschel

Oregon State University
Corvallis, Oregon

Ḱ

Contents

Introduction

Scope and Objective

The intent of this chapter is to provide practical information on how best to ensure that condiments, dressings, and sauces are manufactured to the highest standards of safety and quality. These products are an important part of our diet, as they add an endless variety of tastes, aromas, colors, and textures to otherwise basic foods that may not be of culinary excitement. Could you imagine a hot dog devoid of ketchup or mustard? Or perhaps a bowl of tortilla chips sans the salsa? These products literally provide the spice of life! Open your refrigerator and count the number of them. Undoubtedly they represent the majority of the products you keep there. In the supermarket you see endless shelves with every possible variation and ethnicity of these products represented.

For the new food processor/entrepreneur, these are often the first products that they will manufacture. Often they are encouraged by friends and family who have tasted their homemade product to "go commercial" and they do! Although some give up when they "crunch the numbers," others have the passion, the tenacity, and hopefully a sound business/marketing plan so they are more likely to enjoy success to varying degrees. For the new processor these products are somewhat easier to produce safely, as they are for the most part made shelf-stable by heat pasteurization coupled with being acidified or naturally high in acid. Packing into glass jars simplifies container closure as start-up costs are small compared to double-seam canning operations. The regulatory requirements are significantly less compared to low-acid canned foods (LACFs) or perishable fresh food. From a safety standpoint, they enjoy a somewhat greater degree of safety due to having pH values less than 4.6, effectively precluding the growth of *Clostridium botulinum* and other potential pathogens. These factors coupled with an endless supply of diverse ingredients provide an outlet for culinary creativity that result in the "secret recipe" getting into commercial production. Perhaps the three most popular products we evaluate for compliance to processing regulations are BBQ sauces, salsas, and mustards. Thus we have chosen these three as examples to illustrate various aspects of food processing, preservation, analysis, and safety throughout the chapter.

Definitions

Condiments, dressings, and sauces all to some extent have overlapping characteristics and thus resist having precise definitions. However, Farrell proposed the following definition for condiments: "A condiment shall be a prepared food compound, containing one or more spices, or spice extractives which when added to a food, after it has been served, enhances the flavor of the food."[1]

The author emphasized that it is a food mixture that contains spices and is added to a food after it is served, whereas a sauce may or may not include the use of spices. By contrast, dressings are usually reserved for products that are used in the final preparation of salads. Mayonnaise, salad dressing, and French dressing are also considered to be sauces and are the only products that have a prescribed standard of identity[2] (that indicate specific minimum amount of oil, acids, and egg yolks). Ketchup is the only other condiment that has a standard of identity. It dictates a minimum for the soluble solids content and has an ABC grading classification based on color, flavor, and consistency.[3]

Market Considerations

Anyone who is seriously considering getting into a "for profit" business based on manufacturing condiments, sauces, or dressings needs to have an appropriate market survey. The critical question is, is there a need for your product? If the answer is yes, then the second question is, do you have the resources and skills to make and then market your product? To do such requires a well-thought-out and verified business plan. It is beyond the scope of this chapter to tell you exactly how to answer these questions. Hiring a consultant or working with your local business development agency would be money and time well spent. Universities and community colleges often have programs designed to assist budding food entrepreneurs in various aspects of the food business. It has been the observation of the authors that most food start-up businesses fail within 2 years of launch. Why is this and how can you avoid it?

Major reasons why new food entrepreneurs fail as a business entity may include:

1. Did not establish that there is a sufficient market need for their product
2. Lack of resources and expertise to maintain production at a sustainable/profitable level
3. Failure to effectively market the product to ensure adequate distribution and sales
4. Inconsistent product quality and an inability to resolve production problems
5. Failure to meet regulatory requirements for the production of safe food
6. Unable or not willing to commit the time (24/7) needed to make the venture successful
7. Lost interest, became bored, and the novelty wore off
8. Litigation from product liability
9. Illness and divorce

Avoiding the above situations will likely provide for success as the outlook for the condiment, sauces, and dressing's food sector indicates continued growth, with ethnic/indigenous specialty condiments becoming increasingly popular.[4]

Legal Considerations for the Production of Processed Dressings and Sauces

Regulation of Acidified Foods

Very few if any dressings and sauces fall under the jurisdiction of the low-acid food canning regulations (Code of Federal Regulations (CFR) Title 21, Part 113), which require thermal processes specifically designed to physically destroy spores of *Clostridium botulinum*. Typical food examples would be canned fish and canned low-acid vegetables. Some products will fall under the regulatory umbrella of 21 CFR Part 114—acidified foods where an acid or acid food is intentionally added to lower the pH below 4.6, a value at which *C. botulinum* spores cannot germinate and produce toxin.[5] Other condiments and products that naturally have a pH below 4.6 are classified as acid foods and are not subject to 21 CFR Part 114. Also exempt are carbonated beverages, jams, jellies, and products kept under constant refrigeration. Clarification has been recently proposed by the Food and Drug Administration (FDA) for some of the language in Part 114 addressing dressings and sauces, specifically the statement "acid foods (including such foods as standardized and non-standardized food dressings and condiment sauces) that contain small amounts of low acid food(s) and have a resultant finished equilibrium pH that does not significantly differ from that of the predominant acid or acid food."[6] This can be found in the FDA guidance document 2010 "Guidance for Industry: Acidified Foods"[6] at http://www.fda.gov/Food/GuidanceComplianceRegulatoryInformation/GuidanceDocuments/AcidifiedandLow-AcidCannedFoods/ucm222618.htm, where recommendations are given for what a significant difference in pH may be upon the addition of low-acid ingredients:

- If the equilibrium pH of the predominant acid or acid food is >4.2, then any shift in pH is significant.
- If the equilibrium pH of the predominant acid or acid food is 4.2, then a shift in pH of >0.2 is significant.
- If the equilibrium pH of the predominant acid or acid food is ≥3.8 and ≤4.2, then a shift in pH of >0.3 is significant.
- If the equilibrium pH of the predominant acid or acid food is <3.8, then a shift in pH of >0.4 is significant.

Acid foods that do not exhibit significant changes in pH upon addition of low-acid ingredients are sometimes referred to as formulated acid foods. For more detail, the reader is referred to the FDA acidified foods guidance document.[6]

Good Manufacturing Practices (GMPs), Sanitation Standard Operating Procedures (SSOPs), and Hazard Analysis and Critical Control Points (HACCP) Considerations

Good Manufacturing Practices are published in the Code of Federal Regulations Title 21, Part 110. They are also considered a prerequisite program for other programs, such as HACCP. They should be considered required reading for any managerial employee within any food processing company. The GMPs regulations have evolved over several decades to where they are now fairly broad based with some exceptions, where there are industry-specific GMPs, such as 21 CFR Part 114 (acidified foods), which many sauces and dressings fall under. The Part 114 regulations focus on the need to ensure proper acidification procedures and documentation to prevent the growth of *C. botulinum* in shelf-stable pasteurized foods. The GMPs consist of five subparts that provide definitions relevant to the regulations, requirements for sanitary facilities and buildings, expectations of sanitary design and maintenance of processing equipment, and operation controls to ensure the product is safely processed, such as temperature monitoring and acidification. The last subpart provides definitions for maximum defect levels in certain foods where they are unavoidable and do not prevent a health hazard. For example, in pitted olives an average of 1.3% by count of olives may not have whole pits or fractions thereof exceeding 2 mm. Federal and state inspections are based in part on adherence to the GMPs, and therefore it is prudent to be intimately aware of the requirements. They are in a large part based on common sense and should be intuitive to most. Remember that foods can be declared adulterated if facility conditions are such that the food may become contaminated but may not necessarily be so. Actual contamination does not have to be present for a violation to exist. The GMPs are a preventative set of regulations to ensure conditions are such that they do not have the potential to cause a food safety hazard.

The standard sanitation operating procedures, known as the SSOPs, are essentially your roadmap as to how your operation will adhere to the GMPs. The GMPs specify end results and do not focus on how you go about achieving those results. Written SSOPs are required by regulation only for food operations that are required to have a Hazard Analysis and Critical Control Points (HACCP) plan. However, practically speaking all food processors need to have a plan from which they follow so they can be in compliance with the GMPs. Oftentimes this is a checklist by which specific plant equipment is checked for cleanliness and sanitation on a periodic basis. Other areas routinely checked would be pest control, segregation of nonfood chemicals, water quality, and personal hygiene facilities. Each plant facility would have a different set of SSOPs, but all should be designed to meet and exceed the GMPs.

As mentioned, all operations operating under HACCP are required to have a written SSOP plan. Currently, meat, seafood, and fruit juice processing operations are required to operate under HACCP requirements. Briefly, HACCP is a science-based approach to identify the particular critical points in a food production operation that if not controlled and monitored, may present a food safety hazard. Much has been written on how to put together a HACCP plan, and many universities and private consultants provide training on HACCP. Food companies can develop their own HACCP plans using information such as that provided in *HACCP: A Systematic Approach to Food Safety*.[7]

A compilation of HACCP-related regulations in the United States was published by the Food Processors Institute (2004) and is recommended as a convenient source for reviewing the regulations. Although sauces, dressings, and condiment production do not have a HACCP requirement, manufacturers need to be aware that with the passage of the Food Safety Modernization Act (FSMA) in 2011, almost all food processors will need to have a written food safety plan. The FDA has been charged with establishing science-based standards on which food safety plan requirements will be based. Essentially, facilities will be required to evaluate food hazards, implement preventative controls, monitor and verify the effectiveness of the controls, and maintain records pertaining thereto. Although not specifically called a HACCP plan, it is obvious that most of the HACCP principles will be part of the basis of the required food safety plan. Currently, the FDA is working with academia and industry to establish the specifics of what a plan will require. There will be a public comment period prior to the final rule being in place with an anticipated phase-in date starting 9 months after the comment period. The food safety plan requirement is just one part of the far-reaching FSMA. Besides preventative controls there are provisions for "improving capacity to detect and respond to safety problems" and "improving the safety of imported food." For the text of the act see reference 8. Information workshops regarding the act are being planned by state and federal regulatory entities with "how to be in compliance" with delivery by universities and consultants when the final rule is published. To get ahead of the food safety plan requirement it is suggested that processors go ahead and develop a HACCP plan that will likely satisfy the food safety plan mandate. For some, a HACCP plan is already part of their operations plan, as many buyers already require a HACCP plan from processors as part of Safe Quality Foods (SQF) certification.

Ingredients

Raw Ingredients: Certificates of Acceptance

An important part of processing a safe product is starting with raw commodities that are also safe. Knowing the source of your ingredients is the key to making sure you are getting safe, quality food components. A way to know if a supplier's products have a record of being safe is to ask if they have a certificate of acceptance. These certificates are based on Good Agricultural Practices (GAPs) and Good Handling Practices (GHPs) defined by the U.S. Department of Agriculture (USDA) and other associations with regard to fresh produce.[9] These are voluntary practices that a grower may or may not follow and are intended to protect the agricultural commodity from the field to the packaged product.

GAPs are concerned with the safety of agricultural commodities from the field perspective. These include the growing conditions, such as making sure water sources and fertilizers are not contaminated and that the produce is handled properly during harvest by workers practicing good hygiene. The GHPs are concerned with the storage and packaging of the commodities.[8] The USDA Agricultural Marketing Service offers an audit program that inspects a company's operations and gives a certificate if they pass. Companies that have passed are listed on their website at http://www.ams.usda.gov/AMSv1.0/GAPGHPAuditVerificationProgram. Many other audit services are available through third parties offering certificates as well. If you are getting ingredients directly from a producer, ask to see their certificate. You can also ask a retailer if they have verified that their suppliers have a certificate of acceptance.

Companies can also make agreements with suppliers based on a set of requirements. These requirements can vary from quality parameters to safety parameters to even sustainability practices. The parties can set predetermined limits for various properties, for example, microbial counts, pesticide residues, and chemical properties such as pH and sugar content.[10] The supplier provides to the purchasing company a certificate of specification showing that the agreed upon parameters have been met. The ability to provide such data will depend on the size and productivity of the supplier. Not all suppliers are going to have the resources to conduct extensive testing in-house, in which case they may rely on annual third-party certification for credibility.

Allergen Management

A major concern for any food processor is the presence of food-derived allergens. If a company makes only products that do not contain allergens, then its concerns are limited to trusting that its suppliers do not cross-contaminate raw materials. Companies that choose to produce products that do contain allergens have three very important objectives to keep in mind: protecting their customers, avoiding product recalls, and avoiding lawsuits.

In 2004, FDA passed the Food Allergen Labeling and Consumer Protection Act, or Public Law 108-282. This came about after the FDA conducted a survey and found an alarming number of products that contained allergens yet did not have them listed on the ingredient label. The cause for concern was due to the following statistics:

- Approximately 2% of adults and 5% of children suffer from food allergies.
- 30,000 people are hospitalized annually due to food-induced allergic reactions.

- 150 people die annually due to food allergens.
- 90% of food allergies are related to eight foods.[11]

Proteins found in the offending food cause allergic responses in sensitive individuals. Responses occur due to the interaction of the protein and the human antibody immunoglobulin E, or IgE,[12] and can vary from mild to life threatening. To help allergic consumers avoid these proteins, the FDA began requiring specific labeling of certain food allergens. These are commonly referred to as the "big 8" and include peanuts, tree nuts, fish, crustacean shellfish, soybeans, milk, eggs, and wheat. Additionally, the food coloring carmine has recently been added to the list of compounds that require special labeling. Strict adherence to these labeling requirements will help producers avoid having their product considered misbranded by the FDA and avoid product recalls.

Food allergens must be disclosed in the common name that consumers will recognize. For example, if a salad dressing contains the emulsifying agent lecithin, and it is soybean derived, soy must be disclosed on the label. The font used to identify the allergen cannot be any smaller than the font used for the ingredient list. Bolding of the allergens is not required, but can help consumers locate it. There are two different formats for disclosing food allergens on a label. The allergens can be listed immediately following the ingredient list or adjacent to it as its own list. Alternately, the allergen can be placed in parentheses in the ingredient list. Examples:

- Ingredients: Olive oil, vinegar, sugar, salt, spices, lecithin. **Contains: Soy.**
- Ingredients: Olive oil, vinegar, sugar, salt, spices, lecithin (**soy**).

Additionally, it is not sufficient to just identify an allergen by its broad description. For example, stating that a Caesar dressing (which often contains anchovies) contains fish is unacceptable. Anchovy must be stated explicitly. In the case of the food coloring carmine, it must be listed as carmine or cochineal extract rather than a generic inclusion of "artificial color."[13,14] Public Law 108-282 can be viewed at the FDA website listed in the acknowledgments for further guidance.[11]

Strategies for managing allergens in the processing plant will depend on the product lines being manufactured. A processor that makes products that contain and do not contain allergens will need to exercise far greater caution than the processor that has the same allergen in all of its products. To illustrate how a company might manage the former scenario, a hypothetical processor's strategy is presented below.

ABC Salad Dressing Company currently makes two different varieties of salad dressings: balsamic vinaigrette and toasted walnut vinaigrette. To avoid the raw materials for the balsamic vinaigrette becoming cross-contaminated with allergen residues from the walnuts for the toasted walnut vinaigrette, ABC has a special section of the dry storage area reserved for the walnuts. It also has a storage bin that the walnuts are transferred into once the original packaging has been opened. This bin is clearly labeled "Walnuts Only," and it is never used for any other ingredient storage. Some processors may opt to use color-coded bins to distinguish allergen from nonallergen ingredients. Both the original packages and the storage bin are placed in a manner that at no time is it possible for walnuts to spill onto other ingredients. This is important because if a single walnut were to fall into other ingredients, enough allergen residues may be present to cause an allergic reaction even if the walnut is removed.

During production days, ABC is careful to produce and bottle the balsamic vinaigrette first, since it does not contain any food allergens. The toasted walnut vinaigrette is produced and bottled last to avoid any of the allergen residues from contaminating the other product. After the toasted walnut vinaigrette run is complete, all of the equipment is cleaned and sanitized and checked for residue to ensure that future batches will not be contaminated. This includes making sure that the method of cleaning does not further spread allergen residues throughout the plant. ABC uses low-pressure wash systems to avoid this.[15] Additionally, ABC washes the utensils used during production in the dishwasher and then changes the wash water and cleaning solution to avoid contamination during subsequent washings since their dishwasher recycles these.[15]

ABC is considering adding a Caesar dressing to its line of products. This would introduce egg, anchovies, and milk allergens into its plant. Before deciding to follow through with this product, ABC has to consider proper storage and segregation of the new food allergens and possibly adding another production day just for this dressing to ensure cross-contamination does not occur. Complete cleaning, sanitation, and residue checks have to be performed after each run that contains a different allergen.[16]

All of these precautions are necessary because even if an allergen finds its way into a product unintentionally, it is considered misbranded by the FDA and is subject to a regulatory action. Mislabeling of allergen-containing food products can result in a class 1 recall, which is the most serious and is instituted for dangerous or defective products that predictably could cause serious health problems or death.

Allergens must be considered and treated as a hazard, no differently than other hazards considered with Good Manufacturing Practices.[16] Including a warning that the product "may contain [insert allergen]" does not mitigate any liability a manufacturer is subject to if an allergen is found to be in the product and is not listed as an ingredient explicitly. Additionally, no processor wants to be responsible for the illness or possible death of a consumer due to an allergic reaction that could have been avoided with proper practices in place.

A final item to consider as a food processor is whether or not you intend to export your product. Each country has its own allergen labeling requirements, and these must be met before any product can be sold. A common country for U.S. exports is Canada. In addition to the eight allergens that require labeling in the United States, Canada requires special labeling for sesame seeds and mustard seeds or any derivative of them (e.g., prepared mustard). It is highly recommended to either research a country's policy or contact the proper authority for guidance before planning to export any product.

Organic Certification Requirements

If a processor wants to make a product claiming to be organic, it needs to be aware of the specific legal requirements for including this claim on a product and where on the packaging it can be located. Additionally, marketing a product as organic requires a third party approved by the USDA to certify that it does in fact contain what it claims to.

The labeling requirements are different depending on the percentage of organic ingredients in a product. If a product contains 100% to 95% organic ingredients by weight, it is acceptable to include the USDA organic and certifiers seal on the principal display panel.[17] This is the panel that is intended to be viewed first when a consumer looks at the package. These are the only conditions under which the USDA seal can be used, and there are specifications for what the other 5% of a 95% organic product can be. The certifier is required to be identified on the information panel,[17] which is to the right of the principal display panel. It is required that the organic ingredients be identified as such in the ingredient statement. For example, an ingredient statement for an organic salsa may look like this:

> Ingredients: organic tomatoes, organic onions, organic cilantro, organic jalapeno peppers, organic lemon juice, salt.

Notice that salt is not listed as organic. Salt is a compound not available in an organic form and is allowed in products labeled as organic.

A product can state on the principal display panel that it is made with organic ingredients if at least 70% of the ingredients are organic, and they still need to be identified in the ingredient statement. Processors are allowed to identify organic ingredients in the ingredient statement if the product contains less than 70% organic ingredients; however, the word *organic* cannot appear anywhere on the principal display panel.[17] These are just some of the guidelines set forth by the USDA Agricultural Marketing Service. For additional requirements and further information, it is recommended to visit the USDA Agricultural Marketing Service website at http://www.ams.usda.gov/AMSv1.0/getfile?dDocName=STELDEV3004446.

Processing Parameters for Shelf-Stable Sauces and Dressings

Size Reduction and Blanching

Prior to final processing, some preliminary processing of ingredients may be necessary. This may include reducing particle size or a preprocessing thermal treatment. These processes are done to ensure consistent product results as well as reducing the likelihood of undesirable enzymatic reactions.

Making sure that ingredients are of approximately uniform size helps to ensure that each component of the product gets thermally processed equally. For example, if a marinara sauce has diced carrots in it and they are not uniform in size, some of them may be excessively soft while others are still too crisp after processing. It may also be necessary to reduce particle size to accommodate equipment requirements. If components are too large, they could clog piping networks and feeders.

Enzymes can be problematic during and after processing if they are not deactivated first. Enzymes can survive a wide variety of conditions and can result in quality degradation and discoloration of many products. For this reason, it may be necessary to blanch some ingredients before final processing, particularly if there is a time delay between preparing the ingredients for processing and thermal pasteurization.[18] Blanching involves exposing most often a fruit or vegetable to high heat for a short period of time. This is typically, but not always, done in boiling water for a couple of minutes. The blanched ingredient can then be placed in an ice bath, a process called shocking, to restore a firmer texture, or can be continually processed from this point.[18] This process effectively deactivates most enzymes. Blanching may also be done to reduce the time required to get the product to a desired texture at a later time, but does not reduce the processing time for pasteurization.

An example of this would be if a processor wanted to make a grilled vegetable salsa. They may only want to leave the vegetables on a grill long enough to get the markings and flavor but not long enough to cook the vegetable down to the texture desired. In this case, they could blanch the vegetables and then grill.

Acidification

The vast majority of condiments and sauces are acidic in nature with pH values below 4.6. This is achieved by formulating the product with naturally acid foods or through an acidification of low-acid components with either acid or an acid food. Our discussion hereafter will focus on the acidification process and its technology. Previously we discussed the definitions and regulatory aspects, and later we will review acid/pH as a quality control parameter.

From a historical perspective, it is likely that our first "acidified" foods were those that went through a fermentation process mediated by the lactic acid bacteria, with examples being brined green olives, soy sauce, sauerkraut, kimchi, and an array of cultured dairy products and fermented meats. The metabolic activity of these bacteria converted the sugars to organic acids (lactic and acetic acids), thus increasing the acidity, lowering the pH, and removing sugar, with the net result of producing stable acidic food. Similar products can be produced by the direct addition of acid to foods. However, it is most difficult to reproduce the unique sensory characteristics of traditional fermented foods by the simple addition of acid. Nonetheless, direct acidification provides a controllable process, which within its own category has many desirable food products, for example, a bread and butter pickle (acidified) vs. a genuine dill pickle (fermented). Both enjoy wide market appeal.

Most condiments and sauces are acidic due to either the addition/presence of acid food or direct acidification. In formulating these products, ingredients are mixed and the pH is determined to see if it meets specifications. The target pH value is established from a variety of factors. These include the type of acidulate, its sensory contribution, and process authority recommendation as to establishing a process that will prevent pathogens and spoilage microorganisms from growing. Acidification is most often accompanied with a thermal process to ensure the destruction of non-spore-forming bacteria.

The target pH as established by the process authority is the critical control point to ensure that any spores of *Clostridium botulinum* that may be present are inhibited from germination and toxin production. Although the regulations in 21 CFR Part 114 require a final equilibrium pH value below 4.6, most acidified process authority recommendations are usually lower than 4.4, with pH 4.2 most commonly viewed as a maximum pH. Why? An operating pH value of 4.2 or below provides insurance that it is far less likely that a deviation resulting in a product above 4.5 will occur. If a deviation does occur but the pH is still below, the product may not be necessarily destroyed or recalled after a process authority review. Deviations occur for a variety of reasons, such as failure to add sufficient acid and varying buffer capacity of ingredients. Process authority recommendations for a thermal pasteurization are dependent on the product pH. Generally speaking, the amount of thermal input needed decreases with decreasing equilibrium pH,[19] resulting in energy savings as well as less thermal impact on product quality.

Regulations stipulate the pH values be recorded as "equilibrium" pH values. This is straightforward for homogenous liquid products such as tomato sauce, where acid addition is quickly and evenly distributed during mixing. However, in products that include pieces and chunks, such as a salsa, the distribution of acid throughout all parts of the product may take additional time to be achieved. Measuring the pH before packaging is an obvious point in production to take a reading; however, periodic testing of the packaged product over time would be advisable since chemical reactions that occur during pasteurization and storage may cause the pH to change. 21 CFR Part 114 provides guidance on determining the equilibrium pH. Procedures such as complete homogenization by blending are described. Also described are complete procedures for pH determination and the allowable technology to do so. For example, any product above pH 4.0 must be determined with a potentiometric method, whereas a colorimetric method such as using pH paper (not recommended by the authors) is allowable for products 4.0 and below. Furthermore, methods of acidification such as direct batch acidification, blanching in acid solutions, and addition of acid to individual containers are discussed.

Thermal Processing

A thermal process requirement for acidified foods is defined in 21 CFR Part 114 for acidified foods. Further language gives the option of "permitted preservatives may be used to inhibit reproduction of microorganisms of non health significance in lieu of thermal processing."[5] Practically speaking, process authorities for acidified foods are unlikely to recommend the omission of thermal processing when establishing a schedule process for a processor. Products designated as formulated acid foods (small amounts of low-acid ingredients) may or may not receive a thermal process. Typically mustard products are not thermally processed because it is detrimental to the quality and emulsion stability. It was shown by Rhee et al. (2003) that prepared mustard containing acetic acid and not thermally processed was effective in inhibiting food-borne pathogens, including *Escherichia coli* O157:H7, *Listeria monocytogenes*, and *Salmonella typhimurium*.[20] Part of the antimicrobial effect may be attributable to the presence of allyl isothiocyanate, a natural component of mustard seed that has recognized antimicrobial properties.

Nonetheless, thermal processing of acidified/acid foods is a standard recommendation to ensure the destruction of vegetative microbial cells. Heat processing also provides additional benefits, such as a faster equilibrium of solutes throughout the product, destruction of enzymes that may impact product quality, and the exhaustion of oxygen and a resulting headspace vacuum. As mentioned earlier, the lower the pH product, the less thermal input is needed to ensure commercial sterility. Several validated studies exist on acidified foods to guide the establishment of a schedule process. Acidified foods are evaluated for an adequate safe process by FDA review of Form 2541a. Submission of the form is mandatory for acidified foods and can now be submitted online. One requirement for completing FDA Form 2541a to establish a scheduled process is to designate a process establishment source. That usually entails the identifying of the process authority or the citation of a published study. Instructions for filling out Form 2541a can be found at www.fda.gov and searching the "Forms" section for 2541a instructions. Alternatively, the University of Wisconsin has put online an easier reading set of instructions at http://www.foodsafety.wisc.edu/assets/acidified_foods/ Completing%20Form%20FDA%202541a_Aug2010.pdf.

Most often a letter is appended from the process authority describing the product, how it is processed, and a reference source for the thermal process parameters, which include the F value, Z value, and the reference temperature. The FDA acidified foods guidance document provides a list of validated scientific references for the thermal processing of acidified foods. Two references commonly used are those by Breidt et al. (2010)[19] and Pflug (2008).[21] A typical process authority evaluation letter may look as follows:

January 1, 2011
To Whom It May Concern:
I have been in receipt of a thermally processed pickled asparagus product from_____. This product is classified as an acidified food to which acid (vinegar) is added to a low-acid food (asparagus). The product is pasteurized by covering with hot brine (212°F), sealing and placing in a hot water bath (212°F), and holding until the internal jar center temperature reaches at least 165°F for 2 or more minutes. The actual amount

of hold time will vary with container size. The thermal process establishment source is that published by Breidt et al., 2010, *Food Protection Trends* 30:268–272, that provides validated microbial control processes for acidified foods with a pH value of 4.1 or below. A Z value of 19.5°F and a reference temperature of 160°F with an F value of 1.2 minutes is indicated. The processor needs to make sure the interior center of the products stay at or above 165°F for a minimum of 2 minutes after the jars have been sealed, regardless of the size container used. This can be verified by inserting a bimetallic thermometer through a jar lid during the processing cycle and monitoring the internal center temperature. This will serve to establish the hold time at 212°F for each different container. This should be done once per lot, batch, or run.

The maximum equilibrium pH of the product was determined to be: Pickled asparagus 3.8

Therefore, I conclude that these products conform to the regulations and guidelines put forth in 21 CFR Part 114 when the above-described conditions are adhered to. Complete records of the thermal processing parameters and pH values need to be maintained for each batch or lot.

It is important to note that the actual required thermal input is calculated after the product has been sealed and reaches a designated temperature in the coldest part of the container (geometric center). Several techniques can be used to assess the internal center temperature of products. The bimetallic thermometer is suitable for a hot fill and hold process where the product is not subjected to water, whereas sealed thermocouples or wireless temperature probes (Figure 5.1) are appropriate for hot water baths or for a flow-through pasteurization system. The wireless probe continually collects data throughout the process, which can be downloaded later. Its advantage is that the integrity of the seal remains intact. Most process authority recommendations are in excess of the referenced thermal parameters to ensure that critical deviations are less likely to occur.

Quality Control Parameters

Acid/pH   As previously mentioned, maintaining a pH of below 4.6 is essential for pasteurized acidified foods in order to ensure that *Clostridium botulinum* spores will not germinate and produce toxin. Choosing the optimum acidulant to accomplish this will depend on the desired characteristics of the product. The choice of acidulant is often a result of the desirable sensory profile of the product. Common acidulants include acetic, citric, and lactic acids. They are readily available and are cost-effective. There are product situations where an obvious acid presence is undesirable from a sensory

FIGURE 5.1
Wireless Temperature Data Logger Probe

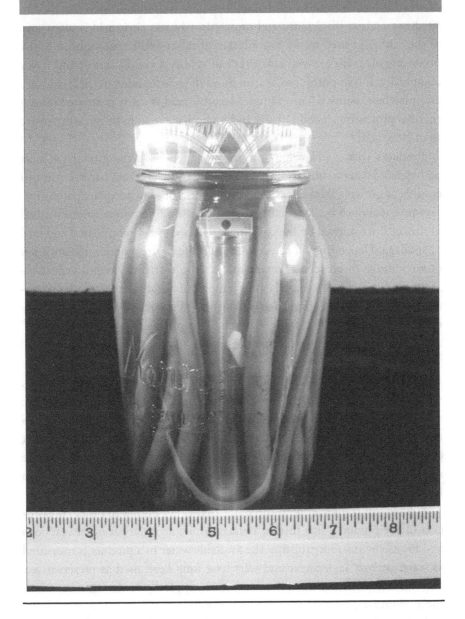

perspective. Glucono-delta-lactone is a food acidulant that hydrolyzes into gluconic acid when added and is less sour than the other acids. Acetic acid, often added as vinegar, has a characteristic aroma, which is well matched to products such as BBQ sauce and mustards, whereas citric acid is often seen as the acidulant used in the formulation of salsas and tomato sauces. Many other organic acids are available that can acidify a condiment or add flavor complexity. It is important to bear in mind that the stability of these can vary and may be expensive. For example, ascorbic acid is easily destroyed during thermal processing,[22] and tartaric and malic acids tend to be more expensive than citric acid.

Products with a pH below 4.6 are protected from the growth of most pathogenic bacteria; however, molds, yeasts, and lactic acid bacteria are still capable of growing in the acidic environment. Thus, an adequate pasteurization protocol must be included to inactivate these microorganisms. Finished products, once opened by the consumer, can become contaminated and lead to spoilage. Depending on the desired shelf life of the product, the addition of an additive to prevent or retard microbial growth may require consideration. Sorbic and benzoic acids are examples of additives that can help control the growth of these spoilage organisms.[23] Product labels will frequently include consumer instructions that state "keep refrigerated after opening," as refrigeration will also help slow the growth of spoilage organisms.

Soluble Solids/°Brix   Soluble solids are solids that can be dissolved in water, with sugar and salt being the most common in processed foods. It is possible to measure the soluble solid content of a product with several types of devices that measure °Brix. This is a measurement typically of sugars in a product expressed as grams of sugar/grams of solution. However, sugars and salts are not the only soluble solids detected in a °Brix reading. Organic acids will also be detected as soluble solids.[24]

Soluble solids interact with the water that is not bound to other constituents of the product. This further reduces available water for other reactions and for use by microorganisms. The available water in a product is measured as water activity (a_w). Sugar and salts have long been used as preservatives, and this is the mechanism of how they work. Typically, as the °Brix increases, the a_w decreases.

Pathogenic organisms do not grow at a_w below 0.85. Most condiments will not be below this value. The protection of the product comes from the low pH coupled with a pasteurization step. In addition, °Brix does not provide a gauge of the safety of the product. °Brix monitoring can, however, monitor the consistency of the quality of a product. For example, if the desired formulation of a salad dressing has a °Brix reading of 13, then deviation from that in test samples may indicate an error in the processing of that batch. Figures 5.2 through 5.6 provide pH, titratable acidity, °Brix, water activity, and sodium content of 21 supermarket purchases of "specialty type" prepared mustards, 20 tomato-based salsas, and 20 tomato-based barbecue sauces. Values are given as the median and range.

Color The color of a product directly affects how consumers perceive it and its freshness, and thus the likelihood of them purchasing and using it.[25] Many factors can affect the color of a product, depending on the nature

FIGURE 5.2
Titratable Ability of Condiments

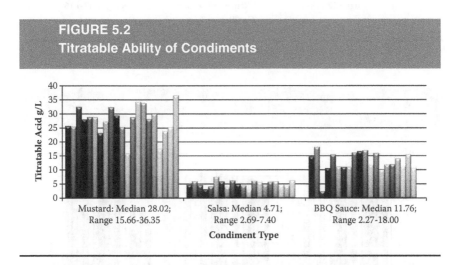

Mustard: Median 28.02; Range 15.66-36.35

Salsa: Median 4.71; Range 2.69-7.40

BBQ Sauce: Median 11.76; Range 2.27-18.00

Condiment Type

FIGURE 5.3
Sodium Content of Condiments

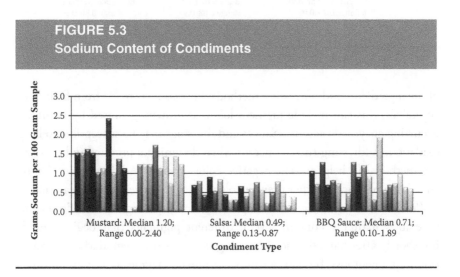

Mustard: Median 1.20; Range 0.00-2.40

Salsa: Median 0.49; Range 0.13-0.87

BBQ Sauce: Median 0.71; Range 0.10-1.89

Condiment Type

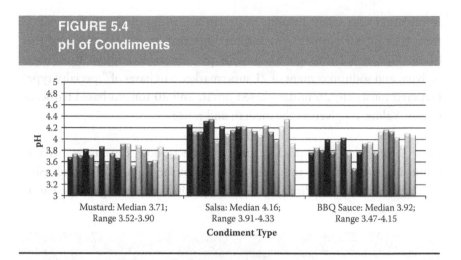

FIGURE 5.4
pH of Condiments

Mustard: Median 3.71;
Range 3.52-3.90

Salsa: Median 4.16;
Range 3.91-4.33

BBQ Sauce: Median 3.92;
Range 3.47-4.15

Condiment Type

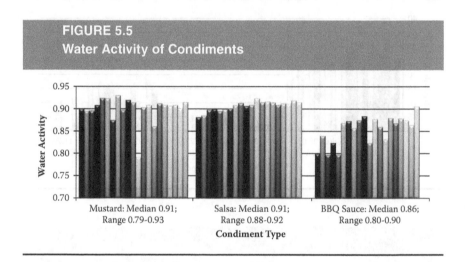

FIGURE 5.5
Water Activity of Condiments

Mustard: Median 0.91;
Range 0.79-0.93

Salsa: Median 0.91;
Range 0.88-0.92

BBQ Sauce: Median 0.86;
Range 0.80-0.90

Condiment Type

of the ingredients and processing methods. For products made from fruits and vegetables, this can include pigment loss and browning caused by naturally occurring enzymatic activity. Sauces and dressings may be susceptible to lipid oxidation. Addressing these concerns when designing a product and processing method can help prevent negative color changes. Additives can also be added to a product to either prevent color loss or improve color appearance.

Products such as salsa will incorporate many fresh ingredients. Fruits and vegetables naturally contain several different pigments that begin degrading shortly after harvest, and processing increases this degradation. In addition to pigment loss, browning can occur due to enzyme activity in produce,

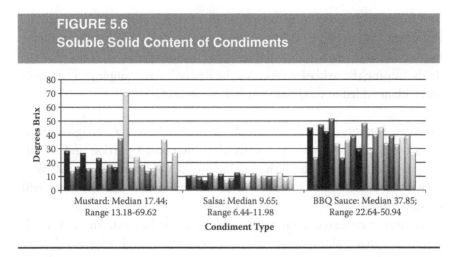

FIGURE 5.6
Soluble Solid Content of Condiments

Degrees Brix

Mustard: Median 17.44;
Range 13.18-69.62

Salsa: Median 9.65;
Range 6.44-11.98

BBQ Sauce: Median 37.85;
Range 22.64-50.94

Condiment Type

particularly if there is any physical damage.[26] The enzymes involved with that process are deactivated during pasteurization. Thermal processing will also cause some pigment loss, as will the addition of acid. Chlorophyll, the green pigment in fruits and vegetables, is pH sensitive and becomes less vivid in acidic conditions.[27] This pigment can be stabilized by the addition of zinc during processing.[28] Anthocyanins, on the other hand, which are blue and red pigments, become more vivid in the presence of acid.[29] The caratenoids, orange and yellow pigments, are less affected by pH changes and more affected by thermal processing.[30] Beware that components in the water supply used in processing may influence or promote color in a product. One notable example is pickled garlic, in which the sulfur components of the garlic can interact with copper ions in the water, imparting a blue-green color to the garlic.

If color loss is an inevitable part of the processing method, additives can be used to improve the appearance of the product. For example, to improve the appearance of a product that has experienced color loss, substances such as turmeric or yellow #5 can be added. Processors should be aware that adding a natural substance such as turmeric to improve color still requires the label to disclose that the product is artificially colored. This disclosure is required if any component is present solely for coloring purposes, regardless if it is a natural or artificial substance.

Lipid oxidation can also negatively impact the color of a product as well as contribute to off-odors. Lipid oxidation occurs when oxygen is introduced to the product. Free radicals can form and attack lipids (fats), resulting in quality deterioration.[31] This process can occur in unopened products. However, it is a significantly more rapid process in opened products. Rates of reactions are slower at colder temperatures; therefore, suggesting that a product be refrigerated after opening can extend the shelf life of the product once the consumer has opened it. Additionally, a processor can add an antioxidant to the product. Free radicals will interact with antioxidants more readily than lipids, causing the oxidative process to be slower.[31] An example of a common antioxidant added to food products is ascorbic acid.

Emulsion Stability   A classic example of an emulsion is a vinaigrette-style salad dressing. An emulsion consists of a water-based portion and an oil-based portion that have been thoroughly mixed and bound in a suspension. When this emulsion is not stable, separation will occur and the layers will become visible.

To create a stable emulsion, an emulsifier needs to be used. An emulsifier is a substance that is able to bind with the water and the oil simultaneously, creating a uniform product. Eggs, which contain lecithin, are a good emulsifier and are often used in products such as mayonnaise. Care must be taken when considering eggs as an emulsifier, and many commercial products use pasteurized egg products to avoid safety concerns. For salad dressings, mustard is frequently used as an emulsifier. Mustard gives the added benefit of being able to introduce additional flavor notes, depending on the variety employed. Many commercial emulsifiers such as lecithin are available. Stabilizers help to keep emulsions together once they have been formed. Examples of these are xanthan and guar gums.

To form a stable emulsion, it is important to incorporate the emulsifying agent into the water-based portion of the formulation first. Oil can then be added slowly to that mixture, ensuring that small oil droplets are introduced and the emulsion can form. Adding the oil too quickly will cause the emulsifier to not be able to interact with each portion adequately, and "breakage" or separation will occur.[32] Be aware that emulsions can be "broken" if products are subjected to extreme environmental conditions (Figure 5.7).

Sensory   Sensory analysis of a product takes into account visible, aroma, taste, flavor, and mouthfeel perceptions as well as other attributes. Having as many positive sensory associations with a product as possible helps consumer perceptions about quality. Understanding what ingredients, processing methods, or packaging can have an impact on the sensory qualities of a specific product is important.

Desired flavor and taste qualities should be kept in mind when choosing an added acidulant. The type of acid used can affect the flavor of the product. For example, tartaric acid may impart a bitter taste, whereas malic acid tends to have a smooth yet tart taste.[33] Citric acid is known for its initial strong tartness.[33]

> **FIGURE 5.7**
> **Emulsion Separation**

For products that contain pieces of vegetables or fruits, excessive thermal processing beyond what is necessary can result in undesirable texture losses in addition to color losses. A novel processing method currently gaining in popularity is high-pressure processing. This technique involves subjecting a product to high hydrostatic pressure, either alone or with temperature increases. This method tends to have less negative impacts on color and texture.[34] High-pressure processing is sufficient for increasing the shelf life of a product and reducing microbial growth in acid and acidified foods. It is, however, an expensive processing method.

Shelf Life Determination   Products that are of low pH have minimal concerns about microorganism growth over time. The shelf life of these products is determined by the manufacturer and is based on what degree of quality deterioration is deemed acceptable.

Periodically testing the product is a way to decide on a period of time that the product should be consumed within for best quality. Evaluating the product at intervals of, for example, 3 months for flavor, appearance, aroma, texture, etc., can help a manufacturer have an idea of how much time can pass and still have the product reach the consumer bearing the sensory qualities intended. It is possible to prolong shelf life with the use of various preservatives and antimicrobial agents.

There are many methods for determining if quality deterioration is detectable. These methods are referred to as discrimination tests and should be conducted blindly, that is, with the persons tasting the samples not knowing which samples are fresh or aged. Examples of discrimination tests are triangle tests and duo-trio tests.[35] A triangle test involves giving test subjects three samples, two of which are the same, for example, from a fresh batch of product, with the third sample coming from an aged batch or vice versa. A duo-trio test involves giving a single sample called a reference sample and then two samples, one of which is from the same batch as the reference and one from a different batch. The subject is then asked to match one of the two to the reference sample. Large companies may have a third party conduct this testing for them; however, many companies conduct this testing in-house by quality control personnel.[35] This can be done to help establish the shelf life of a product, and when conducted on a large enough scale, the data can be analyzed statistically.

Packaging

It is not the intent of this section to provide detailed information on the manufacture and use of food packaging, but rather to illustrate their most common uses with condiments and sauces.

The foremost objective of a package is to provide a secure environment for the food, but at the same time it serves as a platform for consumer information on which a purchase decision is made. By far the most prevalent container for sauces and condiments is a glass closure. It offers the advantage of transparency to showcase product attributes and denote quality by using unique sizes and shapes (Figure 5.8).

Indeed, the cost of a glass closure may eclipse the actual cost of the product within. Relative cost, shipping weight, and potential for breakage

FIGURE 5.8
Glass Containers for Condiments and Sauces

and the accompanying physical hazard are all commonly cited as disadvantages to glass. However, glass remains the standard for the display and safe enclosure of high-quality shelf-stable specialty foods.

Rigid and semirigid plastic food and beverage containers are starting to dominate the packaging industry. An array of materials and the ability to be molded into any shape and color, coupled with being lightweight and cost effective, make this packaging attractive for a variety of food products. Many of our mainstay condiments and dressings are now commonly packaged with these materials, such as ketchup, mayonnaise, and mustard. However, most of the higher-end products still remain in glass, as the clarity and weight are still considered indicators of quality. Many of the food-grade plastics are now of very good transparency and can withstand the rigors of thermal processing. We predict that over the next several decades, plastics will continue to supplant glass, even into the highest echelons of specialty products.

What of the humble double-seam steel can? A mainstay of shelf-stable thermally processed foods, this container is slowly decreasing in use despite innovations such as the two-piece can and more effective surface coatings. Plastics and paperboard materials have entered as alternative materials in the production of low-acid retort processed foods. Examples of ready to "heat and eat" shelf-stable products such as chili and beans and macaroni and cheese have become quite popular. However, the can still has its place in the specialty food arena. Most of our premium fish products still are processed in cans, such as salmon, oysters, and clams. Foie gras is also commonly packaged in cans. Specialty sauces and dressings are rarely found in metal cans with the notable exception of some specialty Mexican-style products such as enchilada sauce.

Labeling

The package provides a platform to provide required information for the consumer to make an informed choice. More and more the public now spends time reading labels to ascertain nutrient content, and the presence or absence of certain food additives. Labeling requirements can be complex, especially in regard to allergens and if organic certification is present. More simple are the requirements for accurate weights and volumes, which are increasingly being notated in both English and metric units. With plastic containers there is labeling of the plastic resin code with an established set of symbols with numeric designations of 1 to 7.[36] The reader is referred to Summers (2007)[37] for an excellent review of the specific requirements for labels to be in compliance in regard to nutrient content, allergen notification, and other required information. It is recommended that processors routinely check the FDA website as labeling requirements frequently are amended as ingredients undergo reclassification, new rules are proposed, and industry guidelines are published. It cannot be stressed enough that label information must be current and accurate. The most frequent cause of class 1 recalls is the failure to properly declare allergens on the label.

Product codes and universal product code (UPC) codes have different functions. The UPC bar code is used to identify products at the point of sale and to facilitate inventory management at the retail level. See http://upcdata.info/ for current information on UPC codes and how to obtain them. They are not designed to satisfy the product code requirement for shelf-stable foods as they do not identify lot number, date of manufacture, and site of manufacture. The latter are important pieces of information to be used to identify products that are being recalled from distribution. FDA and USDA-FSIS require firms that manufacture acidified and low-acid shelf-stable foods to have a procedure in place for recalling foods all the way to the consumer.[38,39] The FDA requires each food container to have a permanent identifying code that conveys information regarding where the product was packed, the identity of the product, and the year, day, and time period when packed (usually a production shift). USDA-FSIS has similar requirements with the exception of the period requirement.

Independent of the UPC and product code is the "use by date" information. Also known as "open dating," where a calendar date is on the product, this serves to provide guidance to retailers as to how long to display the product and for consumers to determine when best to use the product. It is an indicator of quality and not of safety. It is not a regulatory requirement except for infant formula products. How is a "best by," "better by," or "use by" date determined? The processor uses its knowledge and experience with the product to arrive at a date beyond which they believe the quality has diminished to the point that it no longer represents the intent and consumer expectation of the product. Other variations are "use or freeze by" and "refrigerate after opening." The latter is commonly seen on condiments and sauces to prolong shelf life once the product is opened and exposed to the environment. These products are not usually purchased as single serve and used several times, and often are kept in the consumer's refrigerator for extended periods of time. For more information about sell by dates see http://www.fsis.usda.gov/Factsheets/Food_Product_Dating/index.asp.

Summary

Condiments, sauces, and dressings are inherently less likely to present food safety issues than many other food products. Although they typically have water activity values that will support the growth of microorganisms, they have characteristics that provide hurdles to contamination and microbial growth. As processed foods they are packaged in a way that will provide a hermetic seal against environmental contamination. Predominantly they are contained in glass and plastic packages and processed in a manner that provides a shelf-stable product. Preservation is achieved most often from a combination of acid (natural or added), heat pasteurization, and the presence or addition of antimicrobial compounds. The same parameters that ensure safety will by default ensure product quality. You cannot have a quality product without it being safe and protected from spoilage microorganisms.

Condiments, sauces, and dressings are an important and exciting part of our daily menu. They continue to grow in popularity and provide tremendous diversity in their application.[40] Challenges to product safety and quality will come about as more and more new formulations are put on the market, often with obscure ingredients and minimal processing. We are also challenged with providing the consumer with the most accurate label possible.

Failure to declare allergens is often a cause for recalls. Nonetheless, the industry is well poised to address these issues and to continue to move forward in the now firmly rooted specialty food market.

References

1. Farrell, T. 1999. *Spices, condiments, and seasonings.* 2nd ed. Gaithersburg, MD: Aspen Publishers.
2. Code of Federal Regulations Title 21. 21 CFR 169.140 Mayonnaise, 21 CFR 169.150 Salad dressing, and 21 CFR 169.115 French dressing.
3. Code of Federal Regulations Title 21. 21 CFR 155.194 Catsup.
4. Fay, K. 2011. Salad dressings and sauces. Prepared Foods. March 2011. http://www.preparedfoods.com/articles/109078-salad-dressings-and-sauces (accessed March 1, 2011).
5. Code of Federal Regulations Title 21. 21 CFR 114 Acidified foods.
6. FDA. 2010. Guidance for industry: Acidified and low acid canned foods. http://www.fda.gov/Food/GuidanceComplianceRegulatoryInformation/GuidanceDocuments/AcidifiedandLow-AcidCannedFoods/ucm222618.htm.
7. Scott, V., and K.E. Stevenson. 2006. *HACCP: A systematic approach to food safety.* 4th ed. Washington DC: Grocery Manufacturers Association.
8. Food Safety Modernization Act. http://www.gpo.gov/fdsys/pkg/PLAW-111publ353/pdf/PLAW-111publ353.pdf.
9. USDA Agricultural Marketing Service. Grading, verification and certification. http://www.ams.usda.gov/AMSv1.0/ams.fetchTemplateData.do?template=TemplateN&page=GAPGHPAuditVerificationProgram (accessed April 19, 2011).
10. Vasconcellos, J.A. 2004. *Quality assurance for the food industry: A practical approach.* Boca Raton, FL: CRC Press.
11. FDA. 2004. Food Allergen and Consumer Protection Act of 2004. Public Law 108-282. http://www.fda.gov/food/labelingnutrition/FoodAllergensLabeling/GuidanceComplianceRegulatoryInformation/ucm106187.htm.
12. Yunginger, J.W. 1991. Food antigens. In *Food allergy adverse reactions to foods and food additives*, ed. D.D. Metcalfe, H.A. Sampson, and R.A. Simon, 36–51. Cambridge: Blackwell Scientific Publications.
13. Federal Register. 2006. Vol. 71, No. 19. Doc. E6-1104. http://www.fda.gov/ohrms/dockets/98fr/E6-1104.htm.
14. FDA. 2009. Guidance for industry: Cochineal extract and carmine. http://www.fda.gov/ForIndustry/ColorAdditives/Guidance Compliance Regulatory Information/ucm153038.htm.
15. Bagshaw, S. 2009. Choices for cleaning and cross contact. In *Management of food allergens*, ed. J. Coutts and R. Fielder, 114–137. West Sussex: Blackwell Publishing Ltd.
16. Alldrick, A.J. 2009. Risk management-operational considerations. In *Management of food allergens*, ed. J. Coutts and R. Fielder, 102–113. West Sussex: Blackwell Publishing Ltd.
17. USDA Agricultural Marketing Service National Organic Program. 2002. Organic labeling and marketing information. http://www.ams.usda.gov/AMSv1.0/getfile?dDocName=STELDEV3004446.

18. Grandison, A.S. 2006. Postharvest handling and preparation of foods for processing. In *Food processing handbook*, ed. J.G. Brennan, 1–32. Weinheim: Wiley-VCH.

19. Breidt, F., K.P. Sandeep, and F.M. Arritt. 2010. Use of linear models for thermal processing of acidified foods. *Food Prot. Trends* 30:268–272.

20. Rhee, M.-S., S.-Y. Lee, R.H. Dougherty, and D.-H. Kang. 2003. Antimicrobial effects of mustard flour and acetic acid against *Escherichia coli* O157:H7, *Listeria monocytogenes*, and *Salmonella enterica* Serovar Typhimurium. *Appl. Environ. Microbiol.* 69:2959–2963.

21. Pflug, I. 2008. *Microbiology and engineering of sterilization processes*. 13th ed. Minneapolis: Environmental Sterilization Laboratory.

22. Borenstein, B. 1972. Vitamins and amino acids. In *Handbook of food additives*, ed. T.E. Furia, 85–114. Cleveland: CRC Press.

23. Chichester, D.F., and F.W. Tanner Jr. 1972. Antimicrobial food additives. In *Handbook of food additives*, ed. T.E. Furia, 115–184. Cleveland: CRC Press.

24. Sadler, G.D., and P.A. Murphy. 2010. pH and titration. In *Food analysis*. ed. S.S. Nielson, 219–238. 4th ed. New York: Springer.

25. Ragaert, P., W. Verbeke, F. Devlieghere et al. 2004. Consumer perception and choice of minimally processed vegetables and packaged fruits. *Food Qual. Preference* 15:259–270.

26. Busch, J.M. 1999. Enzymic browning in potatoes: A simple assay for a polyphenol oxidase catalysed reaction. *Biochem. Educ.* 27:171–173.

27. Koca, N., F. Karadeniz, and H.S. Burdulu. 2006. Effect of pH on chlorophyll degradation and colour loss in blanched green peas. *Food Chem.* 100:609–615.

28. Ngo, T.X. 2007. Understanding the principles and procedures to retain green and red pigments in thermally processed peels-on pears (*Pyrus communis* L.). PhD diss., Oregon State University.

29. Torskangerpoll, K., and O.M. Andersen. 2005. Colour stability of anthocyanins in aqueous solutions at various pH values. *Food Chem.* 89:427–440.

30. Chen, B.H., H.Y. Peng, and H.E. Chen. 1995. Changes of caratenoids, color and vitamin A contents during processing of carrot juice. *J. Agric. Food. Chem.* 43:1912–1918.

31. McClements, D.J., and E.A. Decker. 2008. Lipids. In *Fennema's food chemistry*, ed. S. Damodaran, K.L. Parkin, and O.R. Fennema, 155–216. 4th ed. Boca Raton, FL: CRC Press.

32. Walstra, P., and T. van Vliet. 2007. Dispersed systems: Basic considerations. In *Fennema's food chemistry*, ed. S. Damodaran, K.L. Parkin, and O.R. Fennema, 783–847. 4th ed. Boca Raton, FL: CRC Press.

33. Gardner, W.H. 1966. *Food acidulants*. New York: Allied Chemical.

34. Nguyen, L., A. Tay, V.M. Bulasubramaniam, et al. 2010. Evaluating the impact of thermal and pressure treatment in preserving textural quality of selected foods. *Food Sci. Technol.* 43:525–534.

35. Lawless, H.T., and H. Heymann. 1999. *Sensory evaluation of food principles and practices*. New York: Springer.

36. SPI resin identification code. http://www.plasticsindustry.org.

37. Summers, J., 2007. *Food labeling compliance review*. 3rd ed. Ames, IA: Blackwell Publishing Ltd.

38. Code of Federal Regulations Title 21. 21 CFR 7 Recall procedures.

39. Code of Federal Regulations Title 9. 9 CFR 318.311 Recall procedures.

40. Nachay, F. 2011. Highflying sauces. *Food Technol.* 65:40–51.

6

Jams, Jellies, and Other Jelly Products

Yanyun Zhao

Oregon State University
Corvallis, Oregon

Contents

Introduction

According to the International Jelly and Preserve Association, approximately 1 billion pounds of fruit spreads are produced annually in the United States. Per capita consumption is approximately 4.4 pounds per year.[1] Annual retail sales for jams, jellies, fruit spreads, and preserves were approximately $632 million. Preserves that consist of jams, jellies, preserves, fruit spreads, and marmalades currently represent 34% total sales in the category. Jams make up 22% sales, with jelly close behind at 21%. Fruit spreads are 17% and marmalades trail with 6%. Sales of specialty sweet spreads, which include conserves, jams, and nut butters, have significantly outpaced mainstream items.[2] According to the *Specialty Food Magazine*'s 2011 State of the Specialty Food Industry Report,[3] specialty conserves, jams, and spreads had a retail sale of $125 million in 2010, an increase of 3.9% from 2008, and the number of new sweet spreads introduced doubled from 2008. Specialty conserves, jams, and nut butters accounted for 12.2% of all food sales in the same food category.[3]

Some of the unique and distinguished characteristics of specialty jellied products may include

- Made of interesting or exotic fruits, or with spices such as ginger root and all spices
- Reduced sugar, or made with no refined sugar at all, thus fruity, less sweetness, and balanced sweet and tart flavor
- Consistency, thick, smooth, and chewy in texture
- Flavor was more full and more interesting
- Vegetable-based jams and preserves, and onion marmalades
- Liquor to be added for flavor, such as scotch with orange marmalade, rum, and champagne
- Spreadable

This chapter discusses the type of jellied products, the unique characteristics of specialty jellied products, the key ingredients and basic procedures for making jellied products, and the specific controls for ensuring quality and microbial safety of jellied products.

Types of Jellied Products

Jams, jellies, preserves, conserves, and marmalades are all fruit-based products that are jellied or thickened. Most are preserved by sugar. Their individual characteristics depend on the kind of fruit used and the way it is prepared, the proportions of different ingredients in the mixture, and the method of cooking. Table 6.1 gives a general definition and description of different jellied products.

Characteristics of Specialty Jellied Products

Specialty jams and jellies are perceived and used as self-indulgent luxuries by consumers. Consumers are willing to pay a higher price for them if they perceive the product to possess the exceptional characteristics that appeal to them. Packaging combined with price is the primary tool that consumers used to judge these products as gourmet, giftable, and otherwise special. According to Denis,[2] some of the unique characteristics of specialty jellied products may include the following:

- Flavors from authenticity of fruit and wild flavors like passion fruit or quince, or combinations like loquat and mint. Rose hip berries, known for subtle sweetness and tanginess, make a jam that has a distinct texture and color. Fig spread provides the fiber and energy with enhanced health benefits.
- Blending fruits to create exotic and fun combination of flavors or adding nuts to old-time favorites for new dimensions in texture. For example, almonds are mixed in peach and almond preserves, pecans are exploding with fruits in Texas, and hazelnuts add a subtle crunch to Oregon pears.
- Fruity flavors plus subtle heat to make more intensive and interesting products. For example, pepper jellies ranging from apricot to pineapple to garlic to mango are whisked into coleslaw and used as finishing sauces for chicken and fish, and split or dual pepper jelly with two colorful flavors in one jar.
- Untraditional marmalades that are more diverse and exotic. Examples are sweet and savory marmalades, including lemon and caramelized onion and blood orange, other tangy citrus fruits like grapefruit, tropical tart and sweet pink grapefruit tea marmalade, and citrus mixes with pumpkin spices.
- A jam-like condiment to complement cheese. Savory or sweet is the latest trend. Hors d'oeuvre-type condiments for entertaining, such as preserves with pecans to serve over cream cheese, sour cherry with lemon balm preserve to pair with triple crémes or goat cheeses, spread with quince and apple spices to complement Roquefort and other blue

TABLE 6.1
Definition and Description of Different Types of Jellied Products

Type of Jellied Products	Definition	Description
Jams	Thick, sweet spreads made by cooking crushed or chopped fruits with sugar. Jams tend to hold their shape but are generally less firm than jelly.	Standards of identify for jam is (Title 21CFR Part 150) 47 parts by weight of the fruit component to 55 parts of the sugar. The finished soluble solids content of a jam is not less than 65%.
Jellies	Usually made by cooking fruit juice with sugar. A good product is clear and firm enough to hold its shape when turned out of the container, but quivers when the container is moved. When cut, it should be tender yet retain the angle of the cut. Jelly should have a flavorful, fresh, fruity taste.	Standards of identify for jelly (Title 21CFR Part 150) is 45 parts by weight of the fruit juice ingredient to each 55 parts of the sweetener solids (45:55). The finished soluble solids content of a jelly is not less than 65%.
Preserves	Almost identical to a jam but preserves can contain large chunks of fruit or whole fruit. The fruit should be tender and plump. Fruit pieces retain their size and shape.	Same standards of identity as jams.
Fruit butter	A spread that is made by cooking fresh fruit with spices until it becomes thick and smooth.	Standards of identify for fruit butter (Title 21CFR Part 150) is 5 parts by weight of the fruit ingredient to each 2 parts by weight of the nutritive carbohydrate sweeteners (5:2). The finished soluble solids content of fruit butter is not less than 43%.
Converses	Jam-like products that may be made with a combination of fruits. They also contain nuts, raisins, or coconut.	No standard of identity.
Marmalades	Soft fruit jellies containing small pieces of fruit or peel evenly suspended in the transparent jelly. They often contain citrus fruit.	No standard of identity.
Fruit spreads	Products are usually made with fruit juice concentrates or low-calorie sweeteners replacing all or part of the sugar.	Do not fall under the jelly or jam standards of identity, usually contain less than 65 percent Brix. Hence the generic name "fruit spreads."

Source. Modified from the National Center for Home Food Preservation, Making Jams and Jellies, Types of Jellied Products, http://www.uga.edu/nchfp/how/can_07/types_jellied_products.html.[4]

cheeses, spread with white figs, bay leaves, and raisins to match chevres, and one with black cherry and licorice to mellow out Brebis, a strong creamy sheep's milk cheese.

Essential Ingredients for Making Jellied Products

There are four essential ingredients in making jelly-type products: fruit or fruit juice, pectin, acid, and sugar. These ingredients should be well balanced in order to form consistent gels and provide desirable flavor.

Fruit or Fruit Juice

Fruit or fruit juice provides the distinguishing flavor of the jellied products. Fruit also contains the sugar, acid, and pectin that are necessary for gel formation. However, the exact amount of sugar, acid, and pectin varies depending on the type, variety, and maturity of the fruit. Hence, the formulation of jellied products even for the same type of fruit may have to be adjusted if the fruit variety and maturity are different. Top-quality, fully ripened fruits are recommended, as they give full flavor and top quality of the final jellied product.

Pectin

Pectin is required for gel formation. It is a natural compound of fruit. The pectin content of fruits varies depending on the fruit variety and maturity. Strawberries, cherries, blueberries, peaches, apricots, raspberries, blackberries, and pineapples contain low pectin, while apples, gooseberries, plums, grapes, cranberries, quinces, and red currents contain a high amount of pectin. Overripe fruits have a lower pectin content than underripe and ripe fruits. In low-pectin fruit, commercial pectin is usually added for gel formation. Commercial pectin is sold at 100 or 150 grade, which is defined as parts of sugar that 1 part pectin will gel under standard conditions at pH = 3.2–3.5 and sugar = 76–70 °Brix. There are liquid and powder types of pectin. They are not interchangeable. Commercial pectin is also classified as low methoxyl and high methoxyl based on its degree of esterification (DE), a process of converting an acid into an alkyl or aryl derivative.

The low-methoxyl pectin has a DE less than 50% and is used for making jelly products with low or no sugar. Specifically:

- It requires 30–60 ppm Ca^{++}/g pectin to gel.
- Calcium salt must be insoluble at the time of addition, which is pH critical.

- Acid is added at the end.
- The product must be filled into containers immediately.

There are two types of high-methoxyl pectin, rapid set with DE usually >70% and slow set with DE between 50 and 70%. For making jelly products using high-methoxyl pectin:

- Use rapid set with DE of 67–75% when fruit pieces need to be suspended. It sets at 80–85°C with an optimal pH of 3.0 to 3.2 and total soluble solid of 60–65%. The typical usage is 0.8% in jams and 1.0% in jellies by dissolving the pectin below 30% sugar solids. Products should be filled above set temperature.
- For slow set with DE of 58–63%, it sets at 50–60°C with an optimal pH of 2.7–2.9 and total soluble solid of 60–65%. The typical usage is 0.8% in jams and 1% in jellies by dissolving it below 30% sugar. Products should be filled into containers above set temperature.

Acid

Acid stabilizes the relation between pectin and sugar, and balances the sweet and sour flavor of jellied products. It can be from the fruits or needs to be added. Underripe fruits have the highest acid content. Lemon juice or citric acid is usually added into low-acid fruit. Acid is necessary for gel formation to reach a pH below 3.5, and also for the flavor.

Sugar

Sugar preserves jellied products by inhibiting the growth of microorganisms, helps formation of gel, and adds flavor. It is the most important water activity depressor in jellied products. Jellied products usually have a water activity of 0.8 to 0.85, in which bacteria would not be able to grow, but yeast and mold can.

The sugar content of fruits also depends on the type, variety, and maturity of a fruit. Additional sugars, usually sucrose and sometimes corn syrup or honey, are added to make standard jellied products. Table 6.2 gives the proper proportions of added sugar with fruit or fruit juice in jellied products based on the standard of identity. Some basic considerations about use of sugar are described below:[5]

- Measure sugar carefully and do not reduce the amount in the recipe. Beet and cane sugars work equally well.

TABLE 6.2

Proper Proportions of Added Sugar with Fruit or Fruit Juice in Jelly Products Based on the Standard of Identify [a]

	Standard Juice °Brix (% sugar)[c]	Weight Units of Sugar per Weight Unit of Fruit Soluble Solids Needed for a Standard Jelly	Formulations per 100 wt units (kg or lb) of Finished Jelly[b]		
			Fruit Soluble Solids Weight Units	Sugar Added Weight Units[d]	Excess Water from Standard Juice Which Must Be Removed Weight Units[e]
Apricot	13.33	9.17	6.39	58.61	6.57
Blackberry	14.29	8.55	6.81	58.20	5.82
Black raspberry	10.00	12.22	4.92	60.09	9.25
Boysenberry	11.11	11.00	5.42	59.58	8.34
Cherry	10.00	12.22	4.92	60.09	9.25
Crabapple	14.29	8.55	6.81	58.20	5.82
Cranberry	15.38	7.95	7.27	57.74	4.98
Currant	10.53	11.61	5.16	59.85	8.82
Fig	10.53	11.61	5.16	59.85	8.82
Grape (Concord)	18.18	6.72	8.42	56.58	2.88
Grapefruit	14.29	8.55	6.81	58.20	5.82
Gooseberry	9.09	13.44	4.50	60.50	10.01
Guava	8.33	14.67	4.15	60.85	10.65
Loganberry	7.69	15.89	3.85	61.15	11.19
Orange	10.53	11.61	5.16	59.85	8.82
Peach	12.50	9.78	6.03	58.97	7.23
Pineapple	11.76	10.39	5.71	59.29	7.81

(Continued)

TABLE 6.2 (CONTINUED)
Proper Proportions of Added Sugar with Fruit or Fruit Juice in Jelly Products Based on the Standard of Identify[a]

	Standard Juice °Brix (% sugar)[c]	Weight Units of Sugar per Weight Unit of Fruit Soluble Solids Needed for a Standard Jelly	Formulations per 100 wt units (kg or lb) of Finished Jelly[b]		
			Fruit Soluble Solids Weight Units	Sugar Added Weight Units[d]	Excess Water from Standard Juice Which Must Be Removed Weight units[e]
Plum	14.29	8.55	6.81	58.20	5.82
Prickly pear	14.29	8.55	6.81	58.20	5.82
Pomegranate	9.09	13.44	4.50	60.50	10.01
Quince	18.18	6.72	8.42	56.58	2.88
Raspberry	13.13	9.17	6.39	58.61	6.57
Strawberry	10.53	11.61	5.16	59.85	8.82

Source.   Modified from Fruit Jellies Food Processing for Entrepreneurs Series, University of Nebraska–Lincoln Extension Publications G1604, http://www.ianrpubs.unl.edu/pages/publicationD.jsp?publicationId=418.[8]

[a] USDA standards of identity for jellies (ratio of 45 weight units of standard fruit juice to 55 weight units sugar concentrated to yield a 65°Brix (% sugar) finished jelly.

[b] Formulation for approximately 100 units (kg or lbs) of finished jelly. The weight would theoretically be increased by 1.54 × weight of pectin and acid supplements used.

[c] The USDA has established standards which define the percent of naturally occurring fruit sugars in many fruit juices.

[d] Excludes water contained in the sugar source used, if any.

[e] Water from the standard single strength juice in excess of the 35 weight units needed in a finished jelly. More water must be removed if a substandard juice is used or if the sugar, pectin, or other ingredients contain water. Less water is removed if the juice is concentrated or above standard.

- Using brown sugar is not recommended because of the dark color imparted to the finished product. Honey or light corn syrup can be used, but note that liquid ingredients must be adjusted accordingly. These sweeteners will also impart a stronger flavor and color to jellied fruit products. There are a wide variety of sugar substitutes available on the market.
- Sugar substitutes can't substitute for sugar if using regular pectin, but can be used to add sweetness when making jams and jellies with low-methoxyl pectin. Sugar substitutes such as sucralose (Splenda) and saccharin (Sweet-n-Low) tend to hold up well during heating. Do not use aspartame (Equal or Nutrasweet), as the resulting product will be unsatisfactory.

With increased consumer consciousness on health, one of the popular trends in specialty jellied products is reduced sugar or no added sugar. The following approaches may be applied to make these products by following specific recipes:[6]

- Use of special modified pectin. These types of pectin are not the same as regular pectin. They are labeled as "light," "less sugar," or "no sugar." Following the directions on the package is essential. Some products are made with less sugar and some with artificial sweeteners.
- Use of regular pectin with special recipes. These special recipes have been formulated so that no added sugar is needed. However, each package of regular pectin does contain some sugar. Artificial sweetener is often added.
- Follow recipes using gelatin. Some recipes use unflavored gelatin as the thickener for the jam or jelly. Artificial sweetener is often added.
- Long boil. Boiling fruit pulp for extended periods of time will cause a product to thicken and resemble a jam, preserve, or fruit butter. Artificial sweetener may be added. Commonly used artificial sweeteners shown in Table 6.3 are suitable to be used in dietetic jams and jellies. All of them have broad acceptance and proven safety.

Processing Procedures for Making Jellied Products

In general, jam and jelly processing involves preparation of fruit, determination of the amount of sugar and pectin to be added, cooking to reach the final total soluble solid amount, adjusting pH, packaging, and cooling. The exact procedures may vary depending on fruit type and specific formulation. The following sections provide the basic processing procedures for making fruit jams and jellies.

TABLE 6.3

Commercial Artificial Sweeteners with the Approval Year, Sucrose Equivalence, and Uses

Sweetener	Year Approved	Sucrose Equivalence	Uses
Acesulfame K	1988 FDA	200 times sweeter	Sweetener—low sugar jam
Aspartame	1981 FDA	200 times sweeter than 4% sucrose solution	Sweetener in non-sugar jam and jelly Presents some breakdown at high temperatures
Sucralose	1991 Canada 1998 FDA	600 times sweeter	Sweetener—jam and jelly Does not break down Better flavor
Sorbitol	1974 GRAS	60% as sweet	Improve viscosity, humectant, hygroscopicity, sweetener

Source. Adopted from Figuerola, F. E., *Berry Fruit: Value-Added Products for Health Promotion*, CRC Press, Boca Raton, FL, 2007, pp. 367–386.

Making Fruit Jams

Figure 6.1 illustrates the general procedures for making fruit jams using berry jam as an example. The procedures are also described below:

1. Select fruit. The fruit should be mature, top quality, and no defected fruit should be used.
2. Prepare fruit. Fruit should be washed with plenty of water and drained, and undesirable parts should be removed. Sometimes, fruits are peeled or cut and seeds are removed when applicable.
3. Calculate amount of sugar. For making a jam, based on 45 parts by weight of fruit to each 55 parts of sugar or other sweetener solids. The finished soluble solids content of a jam is not less than 65%.
   - First, calculate the total amount of sugar by weighing the fruit puree and measuring the Brix. Sugar from the fruit in grams = (°Brix × 10) × fruit weight in kilograms.
   - Weigh out the sugar into a separate container. Calculate the total amount of sugar in kilograms. Total sugar = sugar from the fruit + sugar to be added.
4. Calculate amount of pectin (100 or 150 grade). 150-grade pectin means that 1 part of pectin will gel with 150 parts of sugar at 68° Brix. If the final Brix is 66, increase the pectin by 5%. If the final Brix is 65, increase the pectin by 10%.

FIGURE 6.1
Flow Diagram of General Procedures for Making Fruit Jams

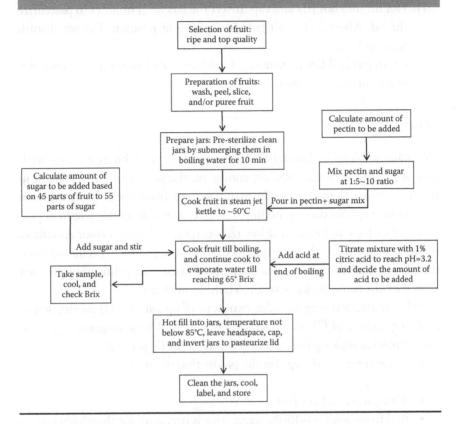

5. Weigh out pectin. Mix pectin with 5 to 10 times its weight of sugar taken from the amount weighed out in step 3. This is to avoid clumping of the pectin when it is added to the liquid fruit.

6. Calculate the total amount of citric acid needed. Weigh out 25 g of fruit and titrate to pH 3.2 with 1% citric acid solution.
   • Citric acid required (g) to titrate 25 g = (titrant volume in ml) × (citric acid concentration/ml titrant).
   • Total citric acid required (g) = amount required for 25 g × 40 × total fruit weight in kilograms.

7. Place fruit or fruit puree in steam jacketed kettle. Heat to approximately 50°C.

8. Slowly pour in pectin-sugar mixture, stirring continuously to avoid clumping.

9. Add remaining sugar while bringing jam up to a boil.

10. Remove a small sample, cool on ice, and check Brix. Continue to boil until Brix reaches 65°. Stir in the acid. If Brix goes above 65°, add water to adjust.

11. Hot fill into hot jars and cap. Invert the jars for 3 minutes to pasteurize the lid. Allow jars to air-cool in an upright position. Do not disturb while gel sets.

12. Clean jars and lids to remove all residues. Label and store in a cool, dry place out of direct light.

Making Fruit Jellies

Jellies are obtained by boiling fruit juices with sugar, with or without addition of pectin and food acids, depending on the pectin and acid content in the fruit juice. They are usually manufactured from juices obtained from a single fruit species only by boiling to extract as much soluble pectin as possible. Jellies have to be clear, shiny, transparent, and with a color specific to the fruit from which they are obtained. Once the product is removed from the glass container, jellies must keep their shape and gelification and not flow, without being sticky or of a too hard a consistency.

Jellies manufacturing includes two steps of operations: (1) production of gelifying juices and (2) manufacturing of jellies.[7] A flow diagram of general procedures for making fruit jellies is described in Figure 6.2.

Here are the general steps for the production of juice:

- Wash, sort, and cut fruits.
- Boil fruits with 50–100% water. This is necessary for the pectin extraction. The boiling time is about 30–60 minutes, and should not be longer in order to avoiding pectin degradation.
- Juice separation. This can be done by a simple drain through a metallic sieve or cloths. In large-scale productions, a hydraulic press is applied.
- Juice clarification. This step is strictly necessary to obtain clear jellies, and can be done by sedimentation during 24 hours or by use of filtration.

For manufacturing jellies, the following steps are usually applied:

- Boil juice to remove about half of the water that has to be evaporated, and then calculate the amount of sugar that needs to be added. The remainder of water is evaporated until the total soluble solids content reaches 65–67% according to the standard identity of jellies.

FIGURE 6.2
Flow Diagram of General Procedures for Making Fruit Jellies

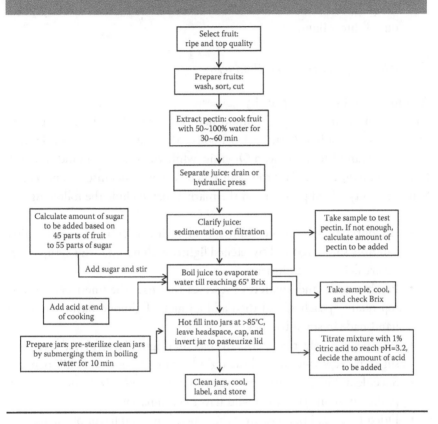

- Remove foam and scum formed to prevent air trapping in the products.
- For juices rich in pectin, gelification occurs without pectin addition. Otherwise, 1–2% powder pectin is added. Same as making jams, pectin is mixed with 10- to 20-fold sugar quantity and is introduced directly in partially evaporated juice and then boiled rapidly up to final point.
- Conduct a rapid test for juice pectin content. For doing so, mix a small sample of juice with an equal volume of 96% alcohol. The apparition of a compact gelatinous precipitate indicates sufficient pectin content for gelification.
- Adjust pH. Product acidity must be brought to about 1% (malic acid) corresponding to pH > 3. When adding acid, it should always be performed at the end of boiling to avoid destroying the acid.
- Boil jellies. Avoid excessively long boiling time, as it may cause pectin degradation.

- Hot fill (not below 85°C) into hot jars and cap. Invert the jars for 3 minutes to pasteurize the lid. It usually takes about 24 hours to allow cooling and product gelification.
- Clean jars to remove all residues. Label and store in a cool, dry place out of direct light.

Examples of Specialty Jellied Products

According to Tanner,[1,9] specialty conserves, jams, and nut butters include gourmet and ethnic items. Common natural standards include the brands of Crofters, Cascadian Farm, Eden Foods, Maranatha, Woodstock Farms, Bionoturae, and Mediterranean Organic, while the specialty brands include Bonne Maman, Dalmatia, Stonewall Kitchen, Jeya, Dickinsons, and Hero. Some specialty jellied products on the market may include the following:

- Unadorned flavor of premium spread with berries and varietal fruits, carefully simmered and sweetened lightly with natural fruit syrup, with no artificial embellishments.
- Peach, pecan, and amaretto preserves that uses the finest pecans and top-quality peaches, and then add a touch of almond flavor. They are handmade in small batches.
- Red pepper jelly that perfectly blends bright flavorful red peppers, sugar, mellow vinegar, and the right kick of jalapeno and chili pepper.
- Succulent kiwi, island pineapple, and the fresh sizzle of jalapenos packed into one jar with a wild flavor combination.
- Dried fig spread that captures the superior natural flavor of the figs.
- Fig and walnut butter that is made from delicious, sweet figs and tasty walnuts. It is a delight for the sophisticated palette and a pantry essential due to its variety of uses.
- Apple jalapeño jelly with roasted apples mixed in fiery jalapeños. It can be simply served with cream cheese and crackers for a quick appetizer or used as a glaze on chicken or pork for an after-work gourmet meal.
- Mandarin pumpkin marmalade with fragrant mandarin and earthy pumpkin marry into a blissful combination of texture and silkiness.
- Sour cherry with lemon balm preserve. The tartness of sour cherries combined with the herbal citrus of lemon balm is fabulous in this whole fruit, not too sweet of a preserve.
- Lime ginger wasabi jams that offer a taste of the tropics with a bit of heat.

- Fruit butters that are rich, dark, and thick with hints of cinnamon and sugar. They rediscover the robust flavor and buttery smooth texture of an American folk classic.

Packaging of Jellied Products

Packaging serves as not only protection from environment contamination and physical damage, but also a marketing tool. Criteria for packaging of jellied products generally include

- Transparent for allowing consumers to view the product
- Barrier to air and water for longer shelf-life
- Resealability for easy consumer handling
- Sturdy so not easily broken

Different types of containers have been used to pack jellied products, with glass jars as the leading and desired one due to their transparency and high water and gas barriers. However, glass has poor physical durability, heavy weight, and relatively high cost. Lately, heat-resistant plastic containers and tubs have been used for jellied products. Examples of thermal-resistant plastics include polyethylene (PET), polypropylene (PP), polystyrene (PS), polyvinyl chloride (PVC), and polytetrafluoroethylene (PTFE). They are lightweight, possess endurance to mechanical strength during shipping and transportation, and are easy to print upon. Flow-type jellied products can be packaged in plastic tubs for providing convenience to consumers. However, when using plastic containers, the presence of air has to be considered due to potential gas permeation into the product that can cause spoiling effect to the product. Spoilage caused by the presence of oxygen can be controlled by the use of a fungistatic preservative, such as potassium sorbate and sodium benzoate. But these chemical preservative are undesirable for specialty jellied products. The use of plastic, either rigid or flexible containers, is normally destined to lower quality or bulk institutional products.

Packaging is extremely important in terms of portraying the gourmet image and inviting sampling on specialty jellied products. According to Uva,[10] comments from consumers about the packaging of specialty jams and jellies include "authentic," "homey," "have a country look," "look home-made," "pretty," "exotic," "very clean, like glass," "smaller," and "wide-mouth jar (to fit spoon)." However, caution should be exercised in fashioning "home-made" packaging to a point where the look might not justify premium pricing, a core value to the appeal of the gourmet jam and jelly market.

Glass jars are most commonly used in specialty jellied products. Patterned glass jars with a unique shape, using colorful labels, and country-style motifs enhance the appeal of a product, especially against strong competition. Additional touches may be added. For example, fiber wrapping has been seen in specialty jellied products to add more appeal, as well as gift packs in hand-crafted wooden containers, including crates, boxes, four-jar wheel arrows, and more. While glass jars add more attractive appearance to consumers, they are more expensive.

Quality Control and Assurance

Quality is commonly defined as "achieving agreed customer expectations or specifications." That is, the customers define the quality criteria needed in a product. To meet these expectations or specifications, the manufacturer puts in a quality control system to ensure that the product meets these criteria on a routine basis.

Quality assurance can be defined as a strategic management function that establishes goals and provides confidence that these measures are being effectively applied. It is the set of systematic and preestablished actions necessary to provide adequate confidence that a product meets the given quality requirement. A quality assurance approach includes the whole production and distribution system, from the suppliers of raw materials, through the internal business management, to the customer. Quality assurance systems should be documented in a simple way to show who has responsibility for doing what and when. Quality assurance focuses on prevention, meaning that action is taken to meet a specification and prevent failures from occurring a second time. This is done by planning, management action, and agreements with key suppliers and other people in the distribution chain.

For jams, jellies, and other jellied products, the causes and possible solutions for problems with jellied fruit products are described in Table 6.4. Some of the important quality factors on the final product and possible procedures for ensuring quality are briefly described below.[5]

- Finishing point: Portion of fruit and added sugar and total soluble solid (TSS) content for meeting the standard of identity of the products.
- Gel setting: Consistency and homogeneity, which are dependent on the gel formation, the pH condition, the amount of pectin, and the relationship between sugar and pectin. If the pH is too low in jams or jellies, it could lead to a very hard gel with a separation of syrup called

TABLE 6.4
Causes and Possible Solutions for Problems with Jellied Fruit Products

Jams and Jellies

Problem	Cause	Prevention
Formation of crystals	1. Excess sugar.	1. Use a tested recipe and measure ingredients precisely
	2. Undissolved sugar sticking to sides of saucepot.	2. Dissolve all sugar as jelly cooks. If necessary, wipe side of pan free of crystals with damp cloth before filling jars.
	3. Tartrate crystals in grape juice.	3. Extract grape juice and allow tartrate crystals to settle out by refrigerating the juice overnight. Strain juice before making jelly.
	4. Mixture cooked too slowly or too long.	4. Cook at a rapid boil. Remove from heat immediately when jellying point is reached. Make small batches at a time; do not double tested recipes.
Bubbles	1. Air became trapped in hot jelly.	1. Remove foam from jelly or jam before filling jars. Ladle or pour jelly quickly into jar. Do not allow jelly or jam to start gelling before jars are filled.
	2. May denote spoilage. If bubbles are moving, do not use.	2. Follow recommended methods for applying lids and processing. (See Mold or Fermentation, below.)
Too soft	1. Overcooking fruit to extract juice.	1. Avoid overcooking as this lowers the jellying capacity of pectin.
	2. Using too much water to extract the juice.	2. Use only the amount of water suggested in the instructions.
	3. Incorrect proportions of sugar and juice.	3. Follow recommended proportions.
	4. Undercooking caused insufficient concentration of sugar.	4. Cook rapidly to jellying point.
	5. Insufficient acid.	5. Lemon juice is sometimes added if the fruit is acid deficient.
	6. Making too large a batch at one time.	6. Use only 4 to 6 cups of juice in each batch of jelly.
	7. Moving product too soon.	7. Do not move jellied products for at least 12 hours after they are made.
	8. Insufficient time before using.	8. Some fruits take up to 2 weeks to set up completely; plum jelly and jellies or jams made from bottled juices may take longer.

(Continued)

TABLE 6.4 (CONTINUED)
Causes and Possible Solutions for Problems with Jellied Fruit Products

Jams and Jellies

Problem	Cause	Prevention
Syneresis or "weeping"	1. Excess acid in juice makes pectin unstable.	1. Maintain proper acidity of juice.
	2. Storage place too warm or storage temperature fluctuated.	2. Store processed jars in a cool, dark, and dry place. Refrigerate after opening.
Darker than normal color	1. Overcooking sugar and juice.	1. Avoid long boiling. Best to make small quantity of jelly and cook rapidly.
	2. Stored too long or at too high of temperature.	2. Store processed jars in a cool, dry, dark place and use within one year. Refrigerate after opening.
Cloudiness	1. Green fruit (starch).	1. Use firm, ripe, or slightly underripe fruit.
	2. Imperfect straining of homemade juice.	2. Do not squeeze juice but let it drip through jelly bag.
	3. Jelly or jam allowed to stand before it was poured into jars, or was poured too slowly.	3. Pour into jars immediately upon reaching gelling point. Work quickly.
Mold or Fermentation (Denotes spoilage; do not use.)	1. Yeasts and mold grow on jelly.	1. Process in a boiling water canner. Test seal before storing. Pre-sterilize jars when processed less than 10 minutes in boiling water.
	2. Imperfect sealing. (Common also with paraffin-covered jellies.)	2. Use new flat lids for each jar and make sure there are no flaws. Pretreat the lids per manufacturer's directions. Use ring bands in good condition—no rust, no dents, no bends. Wipe sealing surface of jar clean after filling, before applying lid.
	3. Improper storage.	3. Store processed jars in a dark, dry, cool place. Refrigerate after opening.
Too stiff or tough	1. Overcooking.	1. Cook jelly mixture to a temperature 8°F higher than the boiling point of water or until it "sheets" from a spoon.
	2. Too much pectin in fruit.	2. Use ripe fruit. Decrease amount if using commercial pectin.
	3. Too little sugar, which requires excessive cooking.	3. When pectin is not added, try ¾ cup sugar to 1 cup juice for most fruits.

Preserves

Problem	Cause	Prevention
Not a characteristic fruit flavor	1. Overcooked or scorched.	1. Should be stirred frequently when mixture begins to thicken to prevent sticking. Cook only to jellying point.
	2. Poor quality fruit used.	2. Select only sound, good-flavored fruit of optimum maturity.
Shriveled product	1. Syrup is too heavy.	1. Follow instructions for the type of fruit being preserved.
Tough product	1. Starting the cooking of fruit in syrup that is too heavy (too much sugar).	1. Cook each fruit according to directions; by evaporation the syrup concentration will gradually increase.
	2. Not plumping fruit properly.	2. Fruit should plump at least 24 hours covered in syrup before canned.
	3. Overcooking.	3. Cook according to directions.
Sticky, gummy product	1. Overcooking.	1. Follow recommended directions for each product. (Cook only until syrup is quite thick and fruit is fairly translucent.)
Darker than normal color	1. Cooking too large of quantities at one time.	1. It is usually best to cook not more than 2 to 4 pounds of prepared fruit at a time.
	2. Cooked too slowly.	2. A better color is usually produced if the product is cooked rapidly.
	3. Overcooked.	3. Cook only until syrup is quite thick and the fruit is fairly translucent.
Loss of color	1. Improper storage.	1. Store processed jars in a dark, dry, cool place.
Mold or fermentation (Denotes spoilage; do not use.)	1. Imperfect sealing.	1. Use new flat lids for each jar and make sure there are no flaws. Pretreat the lids per manufacturer's directions. Use ring bands in good condition—no rust, no dents, no bends. Wipe sealing surface of jar clean after filling, before applying lid.
	2. Yeast or mold growth.	2. Process in a boiling water canner. Test seal before storing. Pre-sterilize jars when processed less than 10 minutes in boiling water.
	3. Improper storage.	3. Store processed jars in a dark, dry, cool place. Refrigerate after opening.

Source. Adapted from the National Center for Home Food Preservation, Causes and Possible Solutions for Problems with Jellied Fruit Products, http://www.uga.edu/nchfp/how/can_07/jellyproblems.html.[11]

syneresis. A higher pH than normal could result in a very soft gel with a very low consistency.
- Acidic taste: Should be well balanced. If too acidic, replace citric acid by tartaric acid.
- Color: A very important quality indicator, especially for berries or other bright-colored fruit jellied products, as it directly attracts consumers. Color is affected by several reactions occurring during processing, including Maillard reaction, ascorbic acid degradation, enzymatic browning, and polymerization of anthocyanins.
- Formation of clots: Probably due to pH being too low or TSS being too high.
- Formation of liquid at the surface: Probably due to too low pH or too low pectin content.
- Crystallization:
  - If liquid forms on the surfaces, the pH is too low, and reducing acid content is necessary.
  - If liquid does not form on the surface, TSS or pH is too high.

Controls for Ensuring Food Safety

Jams and jellies, in general, are safe when they are prepared under controlled conditions. Several factors ensure the food safety and preservation of the products for long shelf-life, including:

- Brix. Adding sugar (65% total soluble solids content) to increase osmotic pressure of the liquid phase, making water unavailable for use by microorganisms.
- Water activity. In food preservation with sugar, water activity is reduced below 0.86, but not below 0.8. This water activity level is sufficient for inhibiting the growth of bacteria and neosmophile yeast, but not preventing the mold attack.
- pH. Between 3.2 and 3.5, a high-acid environment for preventing the growth of most microorganisms.
- Pasteurization. Finished product is pasteurized to kill pathogenic microorganism.

Mold growth may occur on jams and jellies when there is an imperfect seal or too much air space between the jar lid and jellied product. The prevention of mold growth on jams and jellies is important for food safety and economic reasons. New information indicates that molds may be harmful; thus a moldy product should not be eaten. If mold develops, the entire contents of the jar should be thrown away. The best way to avoid mold growth on jams and jellies is to pour the hot product into hot, presterilized jars, leave a headspace for allowing vacuum formation inside the jar after pasteurization and cooling, seal with pretreated two-piece lids, and pasteurize in a hot water bath for the recommended time.

In low-sugar or sugar-free jellied products, food safety depends on the processing of the product. Products have to be thermally processed after formulation and concentration because sugar substitutes do not control water activity, or act as bulking agents. Pasteurization of these products is required by heating at a boiling temperature at atmospheric pressure for about 15 minutes. Normally glass jars have to be cooled after heating in order to avoid the development of thermoduric microorganisms. After opening, the glass jars should be kept in refrigeration for preventing the development and growth of microorganisms.

Summary

Consumer demands on specialty jellied products continuously grow, mainly driven by their distinguished flavor, high quality from the premier quality of fruit, natural and locally grown ingredients, attractive appearance from the color and unique packaging design, etc. With the development of new types of pectin and sugar substitutes, low-sugar or sugar-free jellied products would help to provide healthier and better-quality products to consumers. The future development in specialty jellied products may focus on the use of more varieties of exotic fruit, creative mixtures and combinations of different flavored fruits with spices, herbs, or tea, consistency in texture, attractive natural fruit color, elegant packages, and wide applications that accommodate other food consumption or entertainment.

References

1. International Jelly and Preserve Association. Facts about jams, jelly and preserves. http://www.jelly.org/facts.html (accessed October 1, 2011).
2. Denis, N.P. 2011. Jams, preserves and marmalades: A sweet and savory explosion. *Specialty Food Magazine*. http://www.specialtyfood.com/news-trends/featured-articles/retail-operations/jams-preserves-and-marmalades-a-sweet-and-savory-explosion/ (accessed October 1, 2011).

3. Tanner, R. 2011. *The state of the specialty food industry 2011*. Mintel International and SPINS.
4. National Center for Home Food Preservation. Making jams and jellies, types of jellied products. http://www.uga.edu/nchfp/how/can_07/types_jellied_products.html (accessed October 1, 2011).
5. Figuerola, F.E. 2007. Berry jams and jellies. In *Berry fruit: Value-added products for health promotion*, ed. Y. Zhao, 367–386. Boca Raton, FL: CRC Press, Taylor & Francis Group.
6. Andress, E.L. 2002. Preserving food: Jellied products without added sugar. University of Georgia Cooperative Extension Service Extension Publication FDNS-E-43-12. http://www.fcs.uga.edu/pubs/PDF/FDNS-E-43-12.pdf (accessed October 1, 2011).
7. National Center for Home Food Preservation. Making jams and jellies, extracting juice for jelly. http://www.uga.edu/nchfp/how/can_07/extract_juice.html (accessed October 1, 2011).
8. Fruit Jellies Food Processing for Entrepreneurs Series. University of Nebraska–Lincoln Extension Publications G1604. http://www.ianrpubs.unl.edu/pages/publicationD.jsp?publicationId=418 (accessed October 1, 2011).
9. Tanner, R. 2009. *The state of the specialty food industry 2009*. Mintel International and SPINS.
10. Uva, W.-F. 2005. Marketing specialty jams and jellies to gourmet consumers. Department of Applied Economics and Management, Cornell University. Extension publication. http://marketingpwt.dyson.cornell.edu/SmartMarketing/pdfs/uva7-05.pdf (accessed October 1, 2011).
11. National Center for Home Food Preservation. Causes and possible solutions for problems with jellied fruit products. http://www.uga.edu/nchfp/how/can_07/jellyproblems.html (accessed October 1, 2011).

7

Manufacturing Chocolate for Entrepreneurial Endeavors

Julie Laughter

*Pennsylvania State University
University Park, Pennsylvania*

B. Douglas Brown

*Hershey Foods Corporation
Hershey, Pennsylvania*

Ramaswamy C. Anantheswaran

*Pennsylvania State University
University Park, Pennsylvania*

Contents

Chocolate is a favorite treat for many people because of its sweet taste and smooth texture. Every year, chocolate manufacturers sell ~$13 billion worth of chocolate products in the United States, only accounting for the eighth largest consumption of chocolate worldwide. There are several larger known companies who play a role in creating this billion dollar industry, but most chocolate manufacturing is done by small or medium-sized operations.

In this chapter, ingredients, processing, and specialty items are discussed in hopes of serving as a starting point for those who wish to become an entrepreneur in chocolate manufacturing. Several links and sources are provided to guide entrepreneurs along their path to further understanding and conceptualizing the creative possibilities in chocolate processing.

Standard of Identity

Each country has a legally approved list of ingredients suitable for chocolate production. As noted in Table 7.1, the amount of cocoa solids, liquor, milk solids, cocoa butter, and non-cocoa-based fats is specifically mandated for each country.

The Code of Federal Regulations from the U.S. Food and Drug Administration (FDA) provides more information regarding the standard of identity (SOI) for milk and white chocolate. However, dark chocolate does not have a specified SOI. It is often used interchangeably with semisweet or bittersweet chocolate; so, when making dark chocolate, manufacturers should follow guidelines for the latter two products.

Many countries, including the United States, have regulations regarding the use of optional ingredients. Some of the permitted ingredients include sugar substitute sweeteners, cocoa fat, milk fat, and natural and artificial flavorings. Websites are provided at the end of this chapter with more information about the SOI of various types of chocolate.

Ingredients

Cocoa Ingredients

Cocoa pods provide some of the major components of chocolate, including cocoa powder, cocoa liquor, and cocoa butter. Most cocoa trees are grown in South America, Southeast Asia, and West Africa, with the Ivory Coast, Ghana, and Indonesia as the top three producers of the world's crop. There are four types of cocoa: Criollo, Forastero (most common variety), Trinitario (hybrid of the first two), and Nacional. Each variety is grown in different locations around the world and, consequently, provides its own unique flavor to chocolate.

Extensive processing of cocoa pods must occur in order to obtain the aforementioned chocolate ingredients. Cocoa pods (Figures 7.1 and 7.2) are harvested, fermented, dried, cleaned, and roasted. The shell is removed to obtain cocoa nibs. These nibs are then roasted and pressed to make cocoa

TABLE 7.1
Standard of Identity (SOI) for Milk Chocolate

Regulatory Standards	CODEX		European Union	United States	Mexico [b]			India
Product	Milk Chocolate	Family Milk Chocolate	Milk Chocolate	Milk Chocolate	Milk Chocolate	Chocolate with Milk Content	Chocolate with Low-Fat Milk	Milk Chocolate
Cocoa solids	≥25%	≥20%	≥25%	—	25%	20%	20%	≥2.5%[c]
Free-fat cocoa solids	≥2.5%	≥2.5%	≥2.5%	—	2.5%	2.5%	2.5%	—
Cocoa liquor	—	—	—	≥10%	—	—	—	—
Milk solids	≥12–14%[a]	≥20%[a]	≥14%	≥12%	14%	20%	14%	≥10.5%[a]
Milk fat	≥2.5–3.5%	≥5%	≥3.5%	≥3.39%	2.5%	5%	0.50%	≥2.0%
Cocoa butter	—	—	—	—	20%	17%	20%	—
Total fat	—	—	≥25%	—	—	—	—	≥25%
Maximum vegetable fat (non-cocoa based)	5%	5%	—	5%	5%	5%	5%	Not allowed

Source. Dicolla, C., *Characterization of Heat Resistant Milk Chocolates*, M.S. thesis, Pennsylvania State University, 2009.
[a] Milk solids refer to the addition of milk ingredients in their natural proportions, except that milk fat may be added or removed.
[b] Polyols and artificial sweeteners are within the additives allowed by the Mexican legislation to be added in chocolate.
[c] Based on moisture-free and far-free basis.

FIGURE 7.1
Cocoa Pods and Beans

Source.   Shea Butter-Solidity Trade Blog, http://blog.soliditytrade.com/harvesting-cocoa-beans-to-produce-cocoa-butter.

FIGURE 7.2
Cross Section of Cocoa Pod

Source.   Shea Butter-Solidity Trade Blog, http://blog.soliditytrade.com/harvesting-cocoa-beans-to-produce-cocoa-butter.

liquor, and eventually, after further extraction, cocoa butter and cocoa powder are produced. Figure 7.3 outlines steps involved in processing of cocoa pods to chocolate ingredients.

Cocoa Powder and Liquor   The characteristic color and flavor of milk and dark chocolate partially come from the cocoa ingredients. Cocoa powder is milled cocoa cake. It usually contains 10–12% fat, although there are

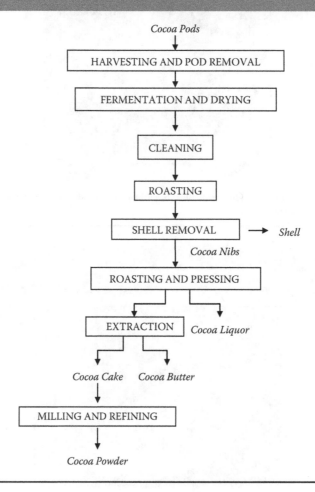

FIGURE 7.3
Processing of Cocoa Pods into Various Chocolate Ingredients

lower-fat options. Some cocoa powders are treated with alkali to develop unique flavors and colors, as depicted in Figure 7.4. Alkalization neutralizes the acidity of cocoa, resulting in less astringent flavors and a variety of colors.

Cocoa liquor is slightly different than cocoa powder in that it contains cocoa solids with cocoa butter in relatively equal proportions. It contains 50–55% fat, 15–20% carbohydrates, 10–15% proteins, and small amounts of moisture, tannins, ash, acids, and caffeine.

Both cocoa powder and liquor should be stored in dry, well-ventilated places to prevent mold and unfavorable microbial growth such as *Salmonella*.

FIGURE 7.4
**Variations of Cocoa Powder Color Resulting from
Different Alkaline Treatments**

Other aromatic foodstuffs should not be in the same warehouse because cocoa is susceptible to picking up their odors.

Cocoa Butter   Cocoa butter is the main fat in chocolate and is responsible for imparting the meltability and smooth mouthfeel characteristic of the candy. It is the most costly ingredient in chocolate, so care must be taken to avoid wasting it. Cocoa butter is obtained by extraction from cocoa liquor. It is primarily composed of 35% palmitic, 23% stearic, and 15% oleic fatty acids.

A polymorphic vegetable fat, cocoa butter can form six crystalline structures with different melting characteristics. Table 7.2 shows the melting profile of the different crystal forms of cocoa butter. Because of the differences in structural orientation, the more heat-stable crystalline forms are capable of packing more tightly together than the less stable forms. During the process of tempering (discussed later), crystal forms V and VI are desired since they are the most heat-stable crystalline forms of cocoa butter. Generally, most manufacturers aim for mostly form V since it is fairly heat stable, provides good snap, and does not induce fat bloom as is thought to be the case for crystal form VI.

Fat Bloom   Fat bloom (Figure 7.5) leads to loss of gloss and a white haze on chocolate products. It results from the migration of liquid fat to the surface of chocolate, where it recrystallizes into larger crystals. Fat bloom occurs through temperature fluctuations from warm environments that melt the fat

TABLE 7.2
Melting Points of Different Polymorphic Forms of Cocoa Butter

Polymorphic Forms	Melting Point Ranges °C (°F)
Form I	16–18 (60.8–64.4)
Form II	21–22 (69.8–71.6)
Form III	25.5 (77.9)
Form IV	27–29 (80.6–84.2)
Form V	34–35 (93.2–95)
Form VI	36 (97.8)

FIGURE 7.5
Dark Chocolate with Fat Bloom

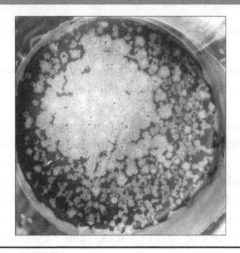

to cooler areas that improperly solidify it. Some sources state that fat bloom also occurs over time when crystal form V transforms to form VI. Form VI packs more tightly than form V, causing the solid fat to contract and squeeze out liquid fat. The migration of soft fats from inclusions such as nuts and nougat into chocolate also leads to fat bloom. In general, storing chocolate at constant temperatures below 20°C (68°F) can delay the appearance of fat bloom.

Cocoa Bean Processing Equipment Entrepreneurs may want to process their own cocoa beans to develop specific flavors. The main pieces of equipment, some of which are seen in Figure 7.6, include a cocoa bean roaster, winnower, grinder, and presser. Beans are harvested, fermented, and dried prior to being shipped. Once at the processing facility, the beans are uniquely

Figure 7.6
Typical Small-Scale Cocoa Bean Processing Equipment

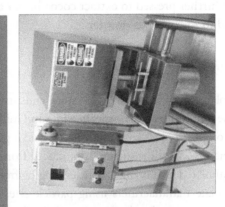

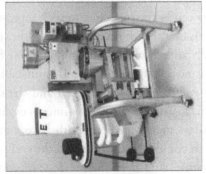

Source.   Courtesy of Bottom Line Technologies.

roasted to develop distinct flavors and colors. Winnowers break the bean shell and vibrate the material over screens to separate the debris from the cocoa nibs. The nibs, which are composed of cocoa solids and cocoa butter, are then ground into a viscous paste known as cocoa liquor. The liquor can be used as is or further pressed to extract cocoa butter as well as cocoa cake, which is ground to make cocoa powder.

Sugars and Artificial Sweeteners

Sucrose   Chocolate usually contains ~50% sugar consisting primarily of sucrose and some lactose from milk components in white and milk chocolate. Sucrose is a disaccharide composed of glucose and fructose, and it gives chocolate its characteristic sweetness. Crystalline sucrose is produced from sugar beet and sugar cane that are granulated into smaller particles ranging from 2.5 mm to 0.1 mm, with most chocolate manufacturers using particle sizes between 0.6 and 1.0 mm. Crystalline sugar can absorb water, especially if stored at higher humidity (>60%). Moisture causes the molecules to stick together and form large lumps of sugar, so sugar is best stored in a dry, low-humidity environment.

Sugar is usually added to chocolate in crystalline form, but it may also be in amorphous, or noncrystalline, form. This unstable structure forms when a sugar solution is rapidly cooled or dried and the sugar molecules are not given enough time to arrange themselves into an organized fashion (i.e., crystalline). Amorphous sugars readily absorb flavors and volatiles from other ingredients, which could be unfavorable in the end product. It is also highly attracted to moisture, and if in the presence of enough water, amorphous sugar can turn into crystalline sucrose, potentially altering processing and texture characteristics of chocolate.

Lactose   Lactose, sugar from cow's milk, is a disaccharide made of glucose and galactose. In crystalline form, it can be added to chocolate to partially replace nonfat milk solids in milk chocolate to reduce costs. Since lactose is one-fifth as sweet as sucrose, it is also used to replace some of the sucrose if the chocolate is too sweet. However, lactose is usually incorporated in chocolate in its amorphous form through milk powders. Compositions of milk powders are discussed later, but in general, whole milk powder contains ~38% amorphous lactose while skim milk powder contains ~52% amorphous lactose. If allowed to crystallize, amorphous lactose can alter the texture of chocolate, imparting grittiness to the finished product.

Lactose undergoes a browning reaction (Maillard reaction) when in the presence of milk proteins and elevated temperatures. This reaction imparts cooked flavors and brown coloration to chocolate. While these changes may be desirable in milk chocolate, care must be taken to avoid this reaction in white chocolate by processing at lower temperatures.

Sugar Alcohols   Sugar alcohols are used to create reduced calorie or sugar-free products. Some common sugar alcohols include xylitol, sorbitol, mannitol, and isomalt. Sugar alcohols are typically not as sweet as sucrose and tend to have a cooling effect in the mouth. Table 7.3 shows the relative percentages of sweetness of common sugar alcohols to that of sucrose. Most sugar alcohols require processing at lower temperatures than sucrose to avoid the formation of gritty lumps. All sugar alcohols should be stored in cool, dry warehouses.

Sugar Bloom   When moisture comes in contact with chocolate, it dissolves sugars. Upon recrystallization of this sugar into large, irregular crystals, a white haze appears on the surface of chocolate. Much like fat bloom, sugar bloom can be delayed and controlled by preventing contact with moisture and avoiding temperature fluctuations.

Milk Ingredients

Milk Fat   Milk fat is responsible for the characteristic smooth texture and "buttery" flavor of milk chocolate and is known to inhibit fat bloom. In comparison with cocoa butter, milk fat contains much shorter fatty acids. When milk fat is combined with cocoa butter, a softer mixture results than if either is used alone, causing some difficulties in creating a solid chocolate. As fat

TABLE 7.3
Sweetness of Sugars and Sugar Alcohols

Sugars and Sugar Alcohols	Percent Sweetness
Sucrose	100
Fructose	130–170
Glucose	60–70
Xylitol	100
Maltitol	75
Sorbitol	50
Mannitol	60
Isomalt	45–65
Erythritol	70

cools, it begins to solidify, but since the milk fat and cocoa butter crystalline forms are structurally different from one another, they do not pack closely together. Usually, lower solidification temperatures are required for chocolates containing both fats. Therefore, using milk fat in conservative amounts is beneficial in creating chocolate with smoother texture and flavor without introducing too many processing difficulties.

The shelf life of milk fat is of major concern for manufacturers because of its ability to oxidize if stored improperly. During oxidation, triglycerides are broken down into shorter-chain free fatty acids, which can impart rancid off-flavors. To counteract this oxidation, milk fat should be stored in chilled refrigerators in oxygen- and moisture-impermeable packaging.

Milk Proteins   Caseins and whey are the main proteins found in milk powders, and they both impart a creamy texture to chocolate. Caseins (~80% of the protein content) are heat stable and have water binding properties, as do whey proteins. Moisture can adversely interact with hygroscopic ingredients that readily absorb water, increasing the viscosity and grittiness of chocolate. Proteins binding this moisture means the water is no longer available to interact with other ingredients, potentially eliminating any unwanted interactions.

Milk Powders   Most milk powders are produced by spray-drying milk. During this process, milk (~50% moisture) is atomized into droplets, which then enter a drying chamber filled with hot air. The milk droplets are quickly dried and collected in filter systems. There are two main categories of milk powders: whole (WMP) and skim milk powder (SMP). Table 7.4 shows the compositions of both powders. WMP particles tend to be larger than SMP because of the encapsulated fat and air vacuoles within them. This means there is less free fat (fat available to reduce viscosity in chocolate) with WMP and more fat may be needed to achieve proper flow properties (discussed later).

TABLE 7.4
Compositions of Milk Powders

Whole Milk Powder	Skim Milk Powder
24.5–27% protein	34–37% protein
36–38.5% amorphous lactose	49.5–52% amorphous lactose
26.6–40% fat	0.6–1.25% fat
5.5–6% ash	8.2–8.6% ash
2–4.5% moisture	3–4% moisture

Since moisture can lead to powder stickiness and caking as well as microbial growth, milk powders should be stored in cool, dry places. Fat oxidation needs to be avoided, so milk powders should always be covered and stored in opaque buckets or bags.

Emulsifiers

Lecithin and polyglycerol polyricinoleate (PGPR) are the most common emulsifiers found in chocolate. They lower viscosity of chocolate, making it easier to mold and enrobe products. Emulsifiers do this by allowing hydrophobic (water-hating) substances like fat to flow more easily by hydrophilic (water-loving) substances like sugar. The hydrophobic portion of the emulsifiers sticks out into the fatty matrix, while the hydrophilic portion of the emulsifiers attaches to the surface of sugars. This bridge between nonmixable substances makes it easier for chocolate to flow.

Lecithin is produced from soya and is added between 0.1 and 0.5% to reduce the viscosity of chocolate. It is much more effective in decreasing viscosity than adding more cocoa butter. However, above 0.5%, lecithin attaches to itself, forms micelles, increases chocolate viscosity, and hinders proper flow.

PGPR is an emulsifier made from castor oil and glycerol. PGPR is usually added between 0.1 and 1.0%. Many manufacturers will use combinations of lecithin and PGPR to improve chocolate flow properties.

Unit Operations for Manufacturing Chocolate

Overview

Dark chocolate contains sugar, cocoa butter, and cocoa powder or liquor. White chocolate also contains sugar and cocoa butter, but instead of cocoa solids, it consists of milk powders and milk constituents. Milk chocolate is in between these two, incorporating both cocoa and milk solids with sugar and cocoa butter. However, all three chocolates are processed similarly, with only slight differences to account for the higher milk portions in milk and white chocolate.

During chocolate manufacturing (Figure 7.7), the main ingredients undergo a reduction in particle size through refining, decrease in viscosity during conching, stabilization with proper tempering of fat crystals, and finally, solidification when molded and cooled. The next section will give an overview of the basic steps of chocolate production as well as information about nontraditional processes used for creating a variety of products.

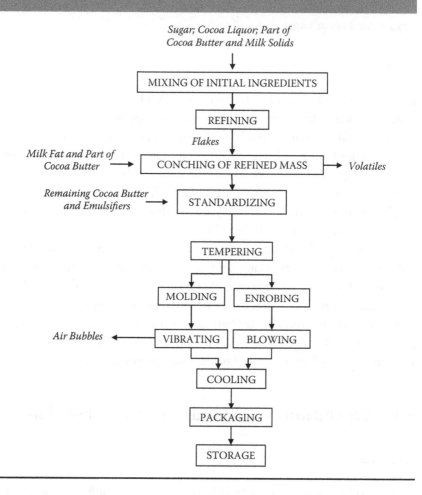

FIGURE 7.7
Process of Manufacturing Milk Chocolate

Refining

Particle size of ingredients impacts texture, taste, and flow properties of chocolate. Most chocolate is refined to particle sizes between 15 and 35 μm, with high-grade chocolate used for making bars below 20 μm and low-grade chocolate used for coating products above 25 μm. During refining, ingredients are broken down to much smaller constituents through shear and compression forces. Initial flavor development also occurs during this process, as ingredients are forced to interact with each other, potentially releasing volatiles.

Prior to refining, sugar, cocoa liquor, milk solids, and part of the cocoa butter are mixed together to coat particles with fat and break apart any agglomerates or large clumps of material stuck together. The refiner is prepared by adjusting the speed of the rolls and pressure between the rolls. The combination of refiner settings determines the final fineness of the particles.

During refining, the mixed paste is first fed into a two-roll or three-roll refiner (Figure 7.8) for prerefining, or the initial breakdown of ingredients. Then, the chocolate is sent back through the three-roll refiner for small-scale operations or onward to a five-roll refiner. A five-roll refiner consists of five cylinders stacked on top of each other. The first cylinder spreads the mixed paste into a thin film. The film is then sent through the remaining rolls. Each roll is faster than the one before it, so chocolate continues moving through the refiner. The rolls proceed to grind the chocolate to the fineness desired. At the end, a knife scrapes the refined chocolate off the roll. A handheld micrometer (Figure 7.9) is often used to measure the particle size of the refined paste after it exits the final roll to determine if any refiner adjustments are necessary.

Fat content plays a vital role in refining. Because of the high roll speeds, fat is essential in sticking particles to the rolls, preventing them from being thrown off the refiner. Therefore, temperature of the rolls (usually around

FIGURE 7.8
Milk Chocolate Running through a Three-Roll Refiner

FIGURE 7.9
Handheld Micrometer Used to Measure Particle Size of Refined Chocolate

40°C or 104°F) must be controlled to ensure fat does not solidify and the film stays on the refiner.

Conching

During conching, chocolate continuously mixes for long periods of time at elevated temperatures. The main objectives of conching are to develop flavors, release moisture, and reduce viscosity of the refined mass. Depending on the temperature, time, and amount of shearing performed, chocolate may leave the conch phase with very different flavors and flow properties. Conching can take anywhere between 3 and 24 hours, depending on desired flavor and flow characteristics. Dark chocolate is conched at temperatures between 60 and 80°C (140 and 176°F), milk chocolate between 50 and 70°C (122 and 158°F), and white chocolate below 50°C (122°F) to avoid discoloration by Maillard browning. However, manufacturers who desire cooked flavors in milk chocolate can use the conching process to encourage the Maillard reaction.

There are three main stages of conching: dry, pasty, and liquid. During the dry phase (Figure 7.10A), the newly refined chocolate is flaky and contains some moisture, which can create undesirable flow properties. The initial shearing action allows for moisture release before the particles are coated in fat, which would make it harder for moisture to escape.

FIGURE 7.10
Three Stages of Conching

  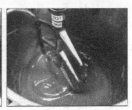

(a) First stage: dry   (b) Second stage: paste   (c) Third stage: liquid

With continuous shearing and increases in temperature, the dry mass becomes pasty when cocoa butter melts and sticks particles together. Balls of chocolate (Figure 7.10B) initially form, and after further shearing, paste is made. Chocolate begins to flow more easily as cocoa butter continues to smear over the various nonfat particles. This coating of fat is very important for the end product as it is the fat that imparts the characteristic smoothness of chocolate as it melts in the mouth.

Finally, chocolate enters the liquid phase (Figure 7.10C) and is standardized through the addition of the remaining fat and emulsifiers to achieve desirable flow properties. In the liquid stage, chocolate is considerably thinner than it was in the previous two phases.

Flow Parameters of Liquid Chocolate

Proper flow parameters are essential in processing chocolate and achieving consumer-acceptable products. There are two main flow characteristics important in chocolate manufacturing: viscosity and yield stress. Viscosity describes a material's resistance to flow, or rather, the thickness of chocolate. Lower viscosities indicate runny materials while higher viscosities are characteristic of thicker material. Yield stress describes the amount of energy required to move chocolate. In other words, yield stress indicates how strongly attracted ingredients are to one another. It describes how quickly chocolate coating sets up. If yield value is high, chocolate sets up readily as seen when depositing baking chips. If yield value is low, chocolate moves more freely and coats products easily.

Several factors affect the viscosity and yield value of chocolate, including shear, particle size, moisture, and the addition of fat and emulsifiers. Chocolate is unique in that it becomes thinner upon being sheared or mixed.

Shearing disrupts interactions between particles, and fat is able to coat particles, decreasing viscosity and yield stress.

As described previously, the refining step reduces the particle size of nonfat solids. Smaller particles increase viscosity and yield stress since they pack closer together than larger, more spherical particles. Therefore, finer chocolate may require more force to move than chocolate with larger particle sizes.

Refining also introduces new surfaces of the nonfat solids, so fat is added during conching to cover them. Fat coats the refined particles and dilutes the chocolate, reducing the viscosity. Yield stress also decreases because particles can no longer interact as strongly with one another.

Most chocolate contains between 25 and 35% fat, with the addition of fat having less of an effect on viscosity reduction past 32%. Many times, emulsifiers (Figure 7.11) are added to decrease yield stress and viscosity by serving as surfactants, compounds that create bridges between hydrophilic and hydrophobic ingredients in chocolate so that they may flow past one another.

Finally, moisture can adversely affect flow properties of chocolate. Water can be either bound or free. Bound moisture does not have much effect on the flow properties of chocolate because it is trapped between or within ingredients. Free moisture, as indicated by its name, can interact with ingredients and cause powder stickiness or sugar crystallization, resulting in formation

FIGURE 7.11
Lecithin (Left) and PGPR (Right) Are Common Emulsifiers Used to Decrease Viscosity and Yield Stress in Chocolate

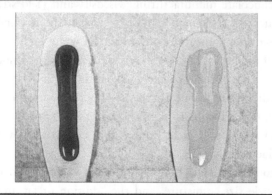

FIGURE 7.12

Cylinder and Bob Used to Measure Viscosity and Yield Stress of Chocolate in a Viscometer

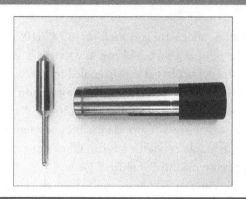

of agglomerated particles or large crystals. Viscosity and yield stress increase, making processing of chocolate difficult. However, most moisture is released during proper conching.

To measure the flow parameters of chocolate, a concentric cylinder viscometer is the most standard instrument used. This technique involves a cylinder (Figure 7.12) filled with conched chocolate. A bob is inserted into the chocolate and rotates at various speeds. The amount of force required by the chocolate to stop the bob from turning is measured. A mathematical model, usually the Casson model, is then used to estimate the viscosity and yield stress. Many viscometers also measure plastic viscosity. This term describes chocolate's flowability once it overcomes its yield stress and begins to move upon shearing. It determines how well the chocolate flows into molds.

Tempering

The process of tempering allows for the formation of the most stable crystalline configurations of fat. This crystal configuration gives chocolate its glossy appearance, good snap, and smooth melting characteristics at body temperature as well as fat bloom resistance and contraction for proper de-molding of finished products. There are three main factors that affect the type, size, and number of fat crystals formed during tempering: temperature, time, and agitation. Cocoa butter crystallizes and melts at certain temperatures, so working within temperature ranges that encourage stable crystal formation is critical. While temperature influences crystal formation, time is also necessary for the crystals to continue growing. Finally, agitation from continuous stirring in the temper bowl distributes the crystals throughout the melted chocolate.

During tempering, chocolate is heated to ~43°C (109.4°F) while continuously stirred in a temper bowl, melting any existing fat crystals. In very controlled conditions, chocolate is slowly cooled to ~26°C (78.8°F) with pre-made seed added along the way to promote the formation of crystal forms IV and V. Lastly, chocolate is heated to 30–33°C (86–91.4°F) so as to melt out form IV, leaving mostly form V. Chocolate is properly tempered when 2–4% of cocoa butter is in stable crystalline form. Chocolate can be tempered by hand or by a tempering machine (Figure 7.13).

White and milk chocolate contain milk fat, which can alter the tempering conditions by interrupting the ability of cocoa butter to crystallize. Therefore, reduced tempering temperatures must be used for fat crystallization to occur in white and milk chocolate. In general, dark chocolate is tempered at slightly higher temperatures (31.1–32.2°C or 88–90°F) because

FIGURE 7.13
Tempering Milk Chocolate

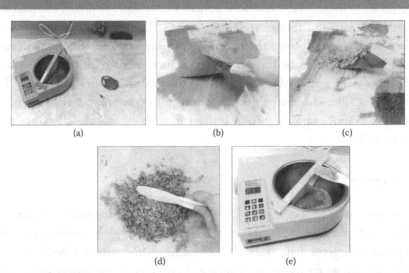

(a)   (b)   (c)

(d)   (e)

(a) Molten conched chocolate is poured into temper bowl with a portion of material placed on a cool, marble slab. (b) Chocolate is spread out and allowed to harden. (c) Hardened chocolate is scraped off the marble slab. (d) Seed is created by finely chopping the scraped chocolate. (e) Conched chocolate is heated to melt out crystals. Chocolate is slowly cooled, during which seed is added to induce proper crystallization of fat.

FIGURE 7.14
Tempered Chocolate Placed into a Small Container
and Put into a Temper Meter Sample Chamber to
Measure the Slope of the Temper Line

of its lack of milk ingredients, while milk and white chocolate are tempered between 28.3 and 31.1°C (83 and 88°F).

A temper meter (Figure 7.14) is used to measure if chocolate is properly tempered. It is composed of a sample chamber, temperature probe, and coolant body. The temper meter monitors the temperature of a small sample of chocolate as it is cooled for a determined amount of time (usually 5–7 minutes) in a controlled fashion. As the temperature is lowered, the fat in chocolate crystallizes, during which latent heat of crystallization is released. This heat is pulled away from the sample chocolate by the coolant body.

If the chocolate is correctly tempered, enough seed is present to properly harden it within the specified measurement timeframe. As the fat crystallizes, latent heat is given off, but the coolant body is capable of taking away the heat at relatively the same rate the fat crystallization produces it. This means the curve of the temper line (Figure 7.15) is close to zero.

Undertempered chocolate does not contain enough seed and fat crystallization is delayed. Once the temperature is low enough, fat eventually crystallizes; however, it does so very rapidly. Latent heat is released so quickly the coolant body cannot take away this heat fast enough. The temperature of the chocolate begins to increase (i.e., the latent heat begins to melt the chocolate). A positive temper curve results.

If there is too much seed, the chocolate sets up too quickly, indicating an overtempered chocolate. The temperature does not have to decrease a lot before the fat crystallizes rapidly, releasing its latent heat. The coolant body offsets this heat; however, the specified amount of time of the measurement is not yet finished. The coolant body continues to take away more heat than what the overtempered chocolate gives off. The temperature of

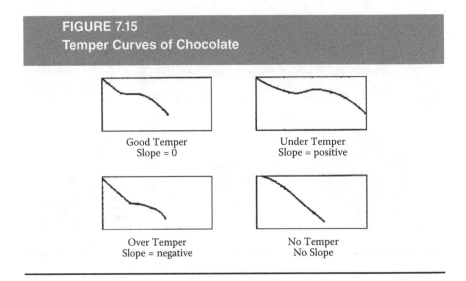

FIGURE 7.15
Temper Curves of Chocolate

Good Temper
Slope = 0

Under Temper
Slope = positive

Over Temper
Slope = negative

No Temper
No Slope

the chocolate sample decreases, and a negative temper curve is seen. Lastly, when no seed crystals are present, the chocolate has no temper, and no slope is seen.

Molding

Once tempered, chocolate is either molded or used for enrobing other products. Molds are usually made of plastic. They are preheated within a few degrees of the tempered chocolate so as not to melt the stable crystals if too hot or set up fat in the wrong crystalline forms if too cold. Chocolate tablets are hand molded as shown in Figure 7.16A or automatically deposited by nozzles into molds at an exact weight. Molded chocolate is then vibrated to remove air bubbles and to distribute the chocolate throughout the mold as depicted in Figure 7.16B. If the air bubbles are not properly removed, the end product has noticeable holes on its surface (Figure 7.17). Depositing chocolate can be difficult if either the viscosity or yield stress is too high. For automated depositors, "strings" of chocolate should not remain between the nozzle and product after depositing.

Enrobing

Tempered chocolate is often used to coat a variety of centers, including nougats, caramels, pretzels, and cookies. Enrobing centers can either be done by hand or through the use of an automated enrober. Figure 7.18 shows a typical enrober in which the centers travel along a continuous

FIGURE 7.16

(a) Tempered Chocolate Is Poured into Molds; (b) Molds Are Vibrated to Eliminate Air Bubbles in the Finished Product

(a)

(b)

FIGURE 7.17

Finished Bar with Holes on the Surface Caused by Air Bubbles Trapped during Molding

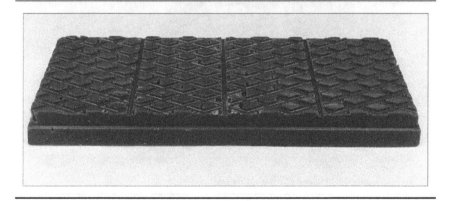

screen belt and pass beneath pouring chocolate that covers the tops and sides of the centers. Excess chocolate is caught in a trough below the belt and recirculated back into the pouring chocolate. A roller beneath the belt coats the bottom of the centers. Products can also be hand enrobed as seen in Figure 7.19.

Once the centers are coated, extra chocolate on the top and sides is blown off by warm air while a knife scrapes off extra chocolate on the bottom. The enrobed products are sent to a cooling tunnel to set the chocolate.

FIGURE 7.18
Enrobing Pretzels with Milk Chocolate

Source.   Courtesy of Hilliard's Chocolate Systems.

FIGURE 7.19
Hand Enrobing of an Inclusion

Source.   Courtesy of Hilliard's Chocolate Systems.

Over time, chocolate collected in the enrobing trough cools down and becomes overtempered. The trough chocolate must be reheated to 40°C (104°F) to melt out all the crystals and tempering is repeated. Care must be taken to ensure enrobing chocolate has the proper yield value. If too low, chocolate will not set up quickly enough on top of the center, resulting in "feet" on the finished product. If too high, chocolate will have difficulty coating the entire center. Figure 7.20 shows a mistake from improper bottoming that can occur on the enrobing production line.

FIGURE 7.20
Improper Bottoming of Product (Right) Compared to
Well-Coated Bottom (Left)

Panning

Panning is a process that covers smaller round items in chocolate or sugar. Open pans or large rotating drums with well-controlled temperature and humidity blowing into them uniformly coat centers. There are three main stages of panning. First, most centers to be panned are pretreated with sugar, starches, gums, and gels to provide them with a protective layer since they hit other centers during panning drum rotation.

During the second stage, chocolate is sprayed onto the inclusions as they tumble over one another. The centers should not be too hot as to untemper the chocolate, but if too cold, the chocolate sets up rapidly and cracks may occur in the finished product. Once the centers are uniformly coated, they are cooled to set the chocolate. During panning, the viscosity and yield value of the chocolate are very important. If too high, chocolate will not move easily over the products, leading to uneven coatings. Conversely, if the flow parameters are too low, chocolate will not stick to the centers, leaving bare spaces. Finally, the centers are covered in a glaze to give the finished product its characteristic gloss and sheen, as illustrated in Figure 7.21.

Chocolate centers can be coated with sugar (Figure 7.22) during a similar panning process. There are two methods of sugar panning. The first involves adding liquid to the rotating chocolate centers. Sugar is then added to "dry out" the products. In another method, sugar is added to the centers in a water-sugar solution, and the products are then dried to evaporate the water.

FIGURE 7.21
Chocolate-Covered Cocoa Nibs

FIGURE 7.22
Sugar-Coated Chocolate Centers

Chocolate Shells

Many chocolate products are hollow inside or have filled centers. To create these candies, chocolate shells are made. Chocolate shells are molded much the same as solid tablets, except instead of completely setting, the molds pass through a cooling tunnel for just enough time to allow the outside of the chocolate to set. The mold is turned upside down, vibrated, and the liquid chocolate that did not harden spreads onto the opposite mold. Both molds are then either cooled to create hollow shells or filled with center material. A spin mold is shown in Figure 7.23. These molds are filled with tempered chocolate, clamped together, rotated to uniformly cover the molds with chocolate, and cooled as illustrated in Figure 7.24. Viscosity and yield stress are important in shell molding in that improper values of either can lead to nonuniform spreading of the chocolate over the molds.

FIGURE 7.23
Molds Used to Make Hollow Shells

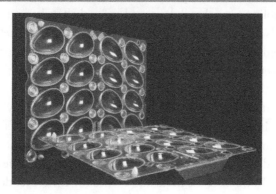

Source.   Courtesy of Brunner Chocolate Moulds.

FIGURE 7.24
(a) Tempered Chocolate Is Deposited into Molds;
(b) Molds Are Attached to Rotary Arms and Spun for
Uniform Coverage; (c) Finished Chocolate Shell Is Made

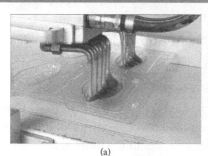

(a)   (b)

(c)

Source.   Courtesy of Brunner Chocolate Moulds.

To fill shells with caramels, pralines, crèmes, etc., the centers should not be hot enough to melt the chocolate. Once filled, a scrapper blade smears chocolate over the center entrance point to finish the shell. Finally, the molds are cooled in a cooling tunnel and deposited onto a belt to be sent to packaging.

Book molds are another device used to make hollow shells. Two shell halves are molded separately. The shell rims are slightly heated so when the two molds meet and close like a book, they stick to one another, forming a hollow shell.

Single-shot depositors are also used to fill chocolate shells with a variety of centers, including caramels, sugar fondant, and peanut butter. This process involves the simultaneous depositing of chocolate and the center. The chocolate is deposited into the nozzle first, and shortly thereafter, the center begins to flow as well. Both enter the mold and eventually the center depositor stops, allowing the chocolate to cover the remainder of the product. The product is made in one moment, hence the single-shot process.

When filling shells with centers using single-shot depositors, it is important that the temperature of the center is not hot enough to melt the chocolate, much like in spin molding. The viscosities of the chocolate and center need to be as close to each other as possible to flow past one another without disturbing the layers of separate materials. Tailing can be an issue if the center does not detach completely from the nozzle and forms a string extending outside the chocolate shell. The center can leak out and stick to packaging or other products.

Cooling

To solidify chocolate, cool air is blown over the product. Importantly, cooling conditions must be very well controlled for two main reasons. One, any moisture in the air can condense on the chocolate, dissolve sugar molecules, and lead to sugar bloom. Therefore, cooling tunnel temperatures must be above the dew point, the temperature at which moisture condensates. Second, fat may crystallize into unstable forms, resulting in fat bloom. Improper fat setting also leads to poor contraction and chocolate becomes difficult to de-mold.

Packaging

Candy is usually an impulse purchase with packaging being the first thing a consumer notices about a product. Creating a package that increases the product's appeal is important to any chocolate manufacturer. A suitable package must provide protection against odors, insects, dirt, and moisture migration. In addition to these protective measures, wrappers also serve as a way to display the product name, "best by" date, coding, and nutritional information.

A common type of package for chocolate bars is foil and paper. Aluminum foil packages provide a moisture and oxygen barrier while also serving as a base to attach paper for advertising the product. Barriers made of polyethylene linings or heat-sealable coatings are often included to protect chocolate from gases and water vapor.

Recently, flow-wrap packaging (Figure 7.25) has replaced the traditional foil-and-paper package. Flow wrapping is faster, reduces packaging material cost, and allows for the continuous production of individually wrapped snacks. The wrapping material is formed into a tube, and the finished product is inserted into it. The tube is then cut and heat-sealed. Twist wrapping (Figure 7.26) is also a popular choice for confectionery products.

There is a multitude of packaging equipment, one of which is shown in Figure 7.27. Depending on the type, amount, and shelf life needs of the finished product, chocolate manufacturers should find the appropriate packaging material and method. Commonly, packages are sealed using heat-seal wraps in which heated clamped jaws seal the edges of the packages. Cold-seal wraps are another option and are meant to give the same result as heat-seal wraps. In this process, the wrapping is coated with latex, pressed against itself, and hermetically sealed.

Storage

Chocolate is stored in dry, dark, cool places with temperatures between 15 and 21°C (59 and 69.8°F) and low relative humidity (below 50%) to avoid sugar bloom, fat bloom, and rancidity. It should not be stored with other

FIGURE 7.25
Example of Flow Wrapping

Source.   Courtesy of Multifilm Packaging Co.

FIGURE 7.26
Example of Twist Wrapping

Source.   Courtesy of Multifilm Packaging Co.

FIGURE 7.27
High-Performance Flow-Wrapping Machine

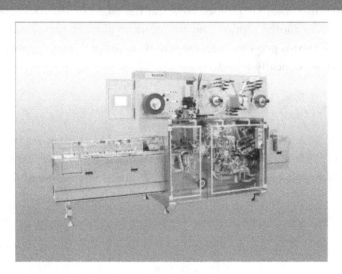

Source.   Courtesy of Bosch Packaging Technology.

foods to avoid aroma absorption. It also needs to be shielded from light and air, as these can stimulate fat oxidation, especially in white chocolate, which does not contain the natural antioxidants found in the cocoa content of milk and dark chocolate. Wrapped chocolate products are usually placed in cardboard boxes to efficiently pack and protect them from breaking.

Standard Chocolate Formulas

While there are many avenues for creativity in chocolate manufacturing, standard formulas can provide a starting point for entrepreneurs. Table 7.5 provides ranges for ingredients specific to dark, milk, and white chocolates.

Food Safety

Chocolate manufacturers should have well defined Good Manufacturing Practices (GMPs) for their processing facilities. Some examples of GMPs include personal hygiene of workers, pest control in the plant, and minimization of ingredient exposure to improper storage conditions. In addition to GMPs, manufacturers can choose to implement a Hazard Analysis and Critical Control Points (HACCP) plan for their processing facility. HACCP allows for the assessment of potential food safety hazards, such as those from physical, chemical, or microbiological dangers.

Processing Hazards

Physical Hazards   There are many potential hazards manufacturers should know of when processing chocolate. Physical hazards include those from incoming raw materials and manufacturing equipment. All incoming materials should be checked for foreign materials, such as metals, wood, or any other contaminant that could enter the production line. Equipment needs to be inspected for any malfunctioning parts on a regular basis.

Chemical Hazards   In addition to physical hazards, chemical dangers can occur preprocessing as well as during the manufacturing of chocolate.

TABLE 7.5
Ingredient Ranges for Different Chocolates

Dark Chocolate	Milk Chocolate	White Chocolate
45–90% cocoa material	15–30% cocoa material	35–50% sugar
15–55% sugar	30–50% sugar	15–25% nonfat milk solids
0–10% cocoa butter	10–20% nonfat milk solids	20–35% whole milk powder
0–0.5% lecithin	15–27% whole milk powder	25–35% cocoa butter
	3–8% milk fat	0–0.5% lecithin
	0–0.5% lecithin	

Mycotoxins and aflatoxins, both fungal-generated toxins, can have nega-tive effects on human health, including kidney and liver failure. By testing incoming raw ingredients and routinely checking the processing facility for these toxins, chocolate manufacturers can reduce the chance of either one making their way into the final product.

During production, cross-contamination between cleaning chemicals and chocolate can occur, so proper sanitation protocols are important. Grease used to lubricate machinery may also contaminate chocolate. Importantly, all chemicals in the plant should be properly labeled, sealed, and stored away from food ingredients.

In recent years, awareness about food allergens, another potential chemi-cal hazard, has increased. Chocolate contains several ingredients that may be problematic for someone sensitive to certain food ingredients. Some prod-ucts are filled with various nuts or peanut butter, so labels should indicate if nut-free products were made on the same line as those made with nuts. Berries and other fruit inclusions should also be clearly stated on the label. Wheat and gluten have also been noted as potential food allergy instigators. Chocolates filled with crispy rice, flour, or wheat may present an issue, but there are several gluten-free products available on the market. Soy lecithin, an emulsifier, is problematic for those with soy allergies, and as such, its inclusion in chocolate should be clearly noted on the label. White and milk chocolate contain milk powder, so chocolate manufacturers can offer dark or dairy-free chocolates for lactose-intolerant consumers. The FDA provides more information about proper labeling of chocolate products.

Microbiological Hazards   Lastly, microbial hazards can affect the safety of chocolate products. If not eliminated prior to production, *Salmonella* and other microorganisms can contaminate several pieces of equipment and remain in the end product. *Salmonella* is of particular concern because it can become heat resistant and protected in the fatty matrix of chocolate. It may be present in raw ingredients such as cocoa beans, milk powders, and peanuts.

Shelf Life

Chocolate is a fairly stable product. Consumer acceptability mostly deter-mines the shelf life, with changes in texture, appearance, and taste being the most common reasons for expiration. As discussed previously, both fat and sugar bloom can impart a white haze to chocolate, deeming it unattractive to consumers, even though neither type of bloom is hazardous to human health. Cracks and broken pieces also visually detract from products.

In general, short storage times ensure better-quality products. White chocolate contains the most milk powder, which can oxidize, so it has the shortest shelf life at ~12 months, followed by milk chocolate at 18 months, and dark chocolate at 24 months. To ensure their products are not out of date, chocolate processors should adopt proper storage rotation of first in, first out.

Specialty Items

Aerated Chocolate

As mentioned earlier, cocoa butter is the most expensive ingredient in chocolate. Aerated chocolate offers a unique alternative for reducing the amount of fat in chocolate. This product gives similar melting and flavor characteristics as regular chocolate; however, instead of fat, gas is used to maintain volume. Chocolate is aerated by vacuum or high pressure, both of which use nitrogen to create air pockets in chocolate, as seen in Figure 7.28.

Non-Standard of Identity Chocolate

Non-standard of identity chocolate is used to coat a variety of products such as ice cream, cookies (Figure 7.29), or other candies. Replacing a portion of cocoa butter with a different, usually higher melting point vegetable fat

FIGURE 7.28
Cross Section of Aerated Chocolate

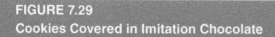

FIGURE 7.29
Cookies Covered in Imitation Chocolate

may provide more appropriate coating properties for some products. Non-standard of identity chocolates should be labeled as such and are generally less costly than regular chocolate.

Diabetic/Diet-Friendly Chocolates

Chocolate usually contains between 30% and 60% sugar, but some consumers want products with much less sucrose for health or diet reasons. No-added-sugar chocolates offer an alternative to regular chocolate. Sugar is replaced, for instance, with sorbitol, fructose, maltitol, or a combination of these. Many low-calorie chocolates contain combinations of maltitol and polydextrose with minimum amounts of fat used. There are several fillings, flavors, and inclusions available to accompany the no-added-sugar chocolate. However, manufacturers must keep in mind that many sugar substitutes, such as sorbitol and fructose, should be processed at temperatures below 50°C (122°F).

Allergen-Free Chocolates

As discussed earlier, allergens have become a major concern for consumers and food processors. Allergen-free products are beginning to emerge in the chocolate industry. Many processors are already offering allergen-free products, including gluten-, nut-, and egg-free chocolates.

Creative Inclusions, Flavors, and Coated Products

Chocolatiers are known for experimenting with new, unique inclusions and fillings. Typical inclusions and coated products include pretzels, nuts, crème, fruits, wafers, nougats, caramels, and toffees. Today, there are chocolate-covered potato chips, jelly beans and rings, coffee beans, edamame, candied ginger, pomegranates, marzipan, dried cherries, and orange slices on the market, much like the assortment of products seen in Figures 7.30 and 7.31.

Chocolate manufacturers also have fun creating chocolates with unique flavors through the addition of spices and artificial or natural flavorings. Chili, curry, cinnamon, ginger, and even red pepper have become popular spices added to chocolate. Other unique flavors include beer, rum, lime, raspberry, sodas, sea salt, and bacon. For entrepreneurs with some imagination, there are countless ways to make chocolate special and distinctive. Figure 7.32 shows some of the possibilities for uniquely flavored products.

Dark Chocolate

Because of its health benefits, dark chocolate has become a consumer favorite. While dark chocolate contains a high amount of fat, approximately a third of the total fat is composed of oleic acids, which are healthy monounsaturated

FIGURE 7.30
Assortment of Chocolate Products

Source. Courtesy of ChocoVision, Inc.

FIGURE 7.31
Chocolate-Covered Jelly Rings

FIGURE 7.32
Unique Flavors Found in Chocolate

fats. Dark chocolate is also high in antioxidants. Small amounts of dark chocolate can help lower blood pressure and LDL (bad) cholesterol. It stimulates production of antidepressants and stimulants such as endorphins, serotonin, and theobromine. Many companies sell dark chocolate with various cocoa contents (Figure 7.33).

Unique Packaging

The first thing a consumer sees when shopping is a product's package. Figure 7.34 shows interesting packaging of chocolate products. Colorful, eye-catching, and bold packaging can enhance consumer acceptability of chocolate products.

FIGURE 7.33
Bars Containing Different Amounts of Cocoa

Figure 7.34
Chocolates with Unique and Colorful Packaging

Personalized Products

Many chocolate companies provide an avenue for their consumers to create personalized products for special occasions and holidays. Consumers can have personal messages printed on wrappers as well as the chocolate products themselves. Some companies offer individualized boxes of assorted chocolates and flavors. There are also mix-and-match programs that allow customers to order a variety of bars to their liking.

Specialty Molds

Molds can also be personalized to reflect seasons, holidays, or special occasions. From Easter bunny molds to heart-shaped chocolates (Figure 7.35), there are countless shapes to create specialized chocolate products for consumers.

Figure 7.35
Heart-Shaped Mold for Specialty Products

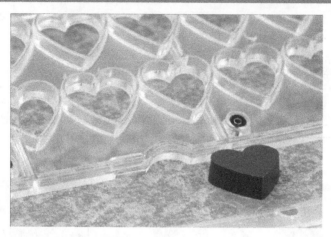

Source. Courtesy of Brunner Chocolate Moulds.

Gourmet Chocolate

Fine, gourmet chocolates contain the highest-grade quality of ingredients. Many gourmet chocolate manufacturers choose not to use artificial flavorings or extracts in their products; instead, they use a variety of spices or natural flavoring to make their chocolates special. Hand-painted designs, sculpted shapes, and eclectic flavors are just some of the products offered by gourmet candy makers.

Organic and Fair Trade Products

Organic foods are one of the increasing trends in the food market, and several chocolate manufacturers are already offering organic products. These items are not genetically modified or processed with any synthetic pesticides or fertilizers, chemical additives, or irradiation. The National Organic Program offers detailed information about how to become organic certified.

Fair trade products are also becoming of interest to some consumers. The movement promotes the rights of marginalized food producers, workers, and farmers. To become fair trade certified, FLO International must audit producers to ensure the fair trade standards are met. Products from fair-trade-certified facilities are marked with the fair trade certification symbol.

Figure 7.36
Variety of Kosher Symbols

Source.   http://qtchaos.tripod.com/ksp.html

Kosher and Halal Certification

Kosher certifications have become more popular for the chocolate industry as products expand into new markets. There are many considerations for entrepreneurs thinking about having kosher-certified products. Products deemed kosher must contain at least one of many symbols, as illustrated in Figure 7.36, indicating the product was made in a kosher-certified facility.

Halal chocolate is another niche area in which an entrepreneur can focus his or her market. In general, no ingredients containing alcohol or animal sources can be used in halal chocolate. Importantly, cocoa liquor, despite its name, does not contain alcohol, so it is suitable for halal chocolate. Like kosher products, halal chocolate must also have a symbol (Figure 7.37) on it indicating it is halal certified. Several companies offer their expertise and services to chocolate manufacturers wanting to become kosher and halal certified.

Future Steps

Marketing

For an entrepreneur wanting to expand his or her business, there are several avenues to choose from when marketing his or her products. First, creating

Figure 7.37
Halal-Certified Symbol

Source.   http://sangiovannipizza.com

innovative, unique, and delicious products is key to any chocolate entrepreneur. Personal company websites provide a way for consumers to browse the various products, prices, and specialty items or services offered. In addition to a personal business webpage, social media provides entrepreneurs a way to market their products to a variety of people and expand their customer base. Asking local shops, like bakeries and florists, to sell their products can help entrepreneurs not only expand their market but also create relationships with other community businesses. Product delivery services are another unique niche for chocolate manufacturers. These systems go beyond selling chocolate products in local stores by bringing them directly to the consumers.

Next Steps

While this chapter provides many aspects of chocolate manufacturing, it is by no means all-inclusive. Several books and resources are provided at the end of this chapter to help entrepreneurs develop their own brand of chocolate. The Institute of Food Technologists (IFT), an international professional organization for those involved in food science and technology, serves as a great starting place to look for information about ingredient and equipment suppliers.

In addition to IFT, entrepreneurs should meet with and join local and national confectionery and food science organizations. Classes on chocolate making are held throughout the world every year and offer a multitude of advice and technical training. Entrepreneurs interested in opening their own chocolate company should take advantage of learning new and traditional processes, networking with fellow chocolatiers, and being creative so they can develop their own unique brand of chocolate. Ultimately, the success of a chocolate company depends upon their product quality, uniqueness, and variety.

Resources

Standard of identity information:
  http://www.accessdata.fda.gov/scripts/cdrh/cfdocs/cfcfr/CFRSearch.cfm?CFRPart=163
  http://www.fda.gov/Food/GuidanceComplianceRegulatoryInformation/GuidanceDocuments/FoodLabelingNutrition/ucm059076.htm
  http://www.hersheys.com/nutrition-professionals/chocolate/composition/definitions/standards-of-Identity.aspx
  www.codexalimentarius.net/download/standards/67/CXS_087e.pdf

Labeling information:
  http://www.fda.gov/Food/FoodSafety/FoodAllergens/default.html

Professional organizations:
  U.S. Small Business Administration, http://www.sba.gov/
  Institute of Food Technologists, http://www.ift.org
  PMCA, http://www.pmca.com/
  National Confectioners Association, http://www.candyusa.com/
  Fine Chocolate Industry Association, http://www.finechocolateindustry.org/
  American Association of Candy Technologists, http://www.aactcandy.org/

Institute of Food Technologists guide to ingredient, equipment, and packaging suppliers:
  http://buyersguide.ift.org/cms/

Other link for ingredient suppliers:
  Food Ingredients website, http://www.food-ingredients.com/

List of equipment suppliers who provided images for this chapter:
  Bottom Line Process Technologies, http://www.cacaocucina.com/

Hilliard's Chocolate System, www.HilliardsChocolate.com

Brunner Chocolate Molds, http://www.hansbrunner.de/index-en.html

Multifilm Packaging Company, http://www.multifilm.com/

Bosch Packaging Technology, http://www.boschpackaging.com/
tevopharm/eng/index.asp

ChocoVision, http://www.chocovision.com/

Micelli Chocolate Mold Co., http://www.micelli.com/

Special thanks to all the companies who provided images for this book chapter as well as Anneliese Luttmann von Ahn for her advice and insight into entrepreneurial endeavors.

Bibliography

Beckett, S.T. 2008. *The science of chocolate*. 2nd ed. Cambridge, UK: Royal Society of Chemistry.

Beckett, S.T. 2009. *Industrial chocolate manufacture and use*. 4th ed. West Sussex, UK: Blackwell Publishing Ltd.

Boyle, T., and T. Moriarty. 2000. *Chocolate passion*. New York: John Wiley & Sons.

Chocolate consumption. In *The world atlas of chocolate*. http://www.sfu.ca/geog351fall03/groups-webpages/gp8/consum/consum.html.

Chocolate tempering. 2008. Presentation at the Confectionery Manufacturing Expo, Brussels, Belgium, June.

Dicolla, C. Characterization of heat resistant milk chocolates. 2009. M.S. Thesis, Pennsylvania State University.

Food allergens. http://www.fda.gov/Food/LabelingNutrition/FoodAllergensLabeling/GuidanceComplianceRegulatoryInformation/ucm106187.htm.

Gott, P.P., and L.F. Van Houten. 1958. *All about candy and chocolate*. Chicago: National Confectioner's Association of the United States.

International Cocoa Organization. http://www.icco.org.

Lebovitz, D. 2004. *The great book of chocolate*. Berkeley, CA: Ten Speed Press.

Minifie, B.W. 1989. *Chocolate, cocoa, and confectionery: Science and technology*. 3rd ed. New York: Van Nostrand Reinhold.

Potter, N.N., and J.H. Hotchkiss. 1998. *Food science*, 464–475. 5th ed. New York: Springer Science + Business Media.

U.S. Dairy Export Council. Milk powders. http://www.usdec.org/home.cfm? navItem Number=82205.

8

Dairy Products

Lisbeth Meunier Goddik and Shawn Fels

Oregon State University
Corvallis, Oregon

Ḱ

Contents

Introduction

Two terms are frequently utilized for specialty dairy products: artisan and farmstead. Neither term is defined in the U.S. Code of Federal Regulations,[1] so the usage is inconsistent. The American Cheese Society is attempting to define specialty, artisan, and farmstead cheeses,[2] and this definition is easily applied to all specialty dairy products:

- Specialty dairy products are defined as products of limited production, with particular attention paid to natural flavor and texture profiles. Specialty dairy products may be made from all types of milk (cow, sheep, goat) and may include flavorings, such as herbs, spices, fruits, and nuts.
- The word *artisan* or *artisanal* implies that a dairy product is produced primarily by hand, in small batches, with particular attention paid to the tradition of the processor's art, and thus using as little mechanization as possible in the production of the cheese.
- Farmstead products must be made with milk from the farmer's own herd on the farm where the animals are raised.

Thus all farmstead products are artisan, but not all artisan products are farmstead.

Fluid milk has been successful as a specialty product in situations where it can be differentiated from commodity milk products. Common examples are unhomogenized fluid milk, organic milk, milk destined for home delivery, and milk from goats, sheep, and even buffalo. The most common specialty fluid milk product is raw milk. Federal regulations in the United States currently restrict interstate sale of raw milk for consumption. States differ in their regulations concerning raw milk sales. Some freely allow sale from licensed dairy farms, others allow limited on-farm sale and cow shares, while many states completely outlaw the sale of raw milk. In some states the regulations make a distinction between the different animals producing milk as well. The state-by-state situation is currently in flux as the issue is debated by state legislators across the United States. Artisan cheeses are the most successful specialty dairy products. The specialty market is growing faster now as many retailers feature local cheeses. American specialty cheeses are competing successfully in global cheese competitions and are now beginning to be exported to other continents. The reason for the success of the specialty cheese sector is the opportunity for value added associated with cheese, along with the lack of competition from the mainstream commodity cheese sector.

As a consequence, the economic outlook for a specialty cheese plant is considered more positive than for specialty fluid milk or yogurt plants.[3]

Butter is made from cream and salt. Whether produced by old-fashioned churn or modern continuous churning, the final products tend to be similar. Yet to succeed with producing and marketing specialty butter, it is essential to differentiate the product from mass-produced alternatives. There are several approaches for accomplishing this. The most common is to produce butter from cultured cream, which is the traditional way of producing butter. The fermentation naturally brings out aromatic compounds, such as diacetyl, the typical butter flavor, which is produced as *Leuconostoc mesenteroides* ssp. *cremoris* metabolizes citrate. In commodity butters, diacetyl is not naturally produced but may be added as a flavor enhancer. Producing cultured butter is one way to differentiate the product; another option is to pick a nontraditional salt such as sea salt or flavored salt. Another approach is through packaging. Unique butter containers help differentiate and elevate the perceived quality of specialty butters.

Specialty fermented fresh products include kefir, yogurts, and crème fraîche. All are relatively simple to produce and lend themselves to being marketed as specialty products. Only yogurt faces significant competition from "mainstream" yogurts. Therefore specialty yogurts must have a unique twist, such as "cream on the top" nonhomogenized yogurt, full-fat yogurt with no stabilizers, or products from specific animals, such as Jersey, goat, or sheep milk yogurt.

Common to all specialty dairy products is the need to tell the story to consumers. Consumers who are willing to pay more for a specialty product want to be assured that the product is produced in a unique and wholesome manner using milk from animals that have been treated correctly. When selling products at farmers' markets there is a direct contact with customers, and it is easy to explain the story behind the product. Wholesale and distribution are channels with limited communication. Processors selling products through these channels must find other ways of communication, and frequently the best option is on the product label. A recent study evaluated factors that impact purchasing decisions for consumers of specialty cheeses.[5] The consumers expressed a strong interest in cheeses made from raw milk, local products, labels with the cheese maker's story, and products that were artisan. Terms such as *farmstead* and *organic* were less important. Perhaps the lack of emphasis on organic relates to people's conviction that local specialty dairy products are already produced without artificial hormones and with respect to animal welfare. Thus the common drivers for organic products might not provide a competitive advantage.

Processing of Specialty Dairy Products

Fluid Milk

Processing of Fluid Milk Upon arrival to the processing plant, raw milk is tested for beta-lactam residues and temperature. The milk must not contain detectable beta-lactams, and the temperature must be below 45°F. The raw milk is stored in a cooled bulk tank until processing. Fluid milk is typically separated into skim and cream in a centrifugal clarifier/separator and restandardized to the target fat content. If vitamins are supplemented, it is commonly done at this point. The milk is then homogenized to prevent creaming. Pasteurization must follow immediately after homogenization to prevent the development of rancid off-flavors resulting from the reduction of fat globule integrity. The pasteurized milk is either packaged immediately or stored in a cooled tank until packaging. Manual packaging is generally not utilized due to the requirement for automatic capping of bottles.

Separation Separators (Figure 8.1) use centrifugal force to separate the lower-density fats in milk from the higher-density water, protein, and minerals. Milk enters as a continuous stream in the bottom of the centrifuge. As

FIGURE 8.1
Milk Separator

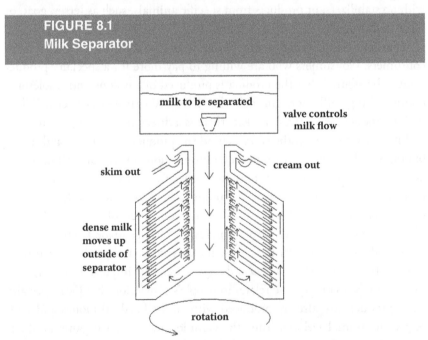

the centrifuge spins, milk moves upward and passes over a series of plates. Between individual plates the heavier skim milk fraction is forced outward by centrifugal force and the lighter cream is pushed inward, thus separating the milk stream into skim milk and cream.

Homogenization   Homogenization is the application of mechanical force to milk to reduce the average milk fat globule size. The reduction in size along with incorporation of caseins in the new fat globule membranes stabilizes the milk fat emulsion and prevents creaming. Specialty milks are frequently not homogenized to differentiate from commodity type milks. Unhomogenized pasteurized milk must be continuously but gently stirred prior to filling to avoid separation of cream. Otherwise the fat content would vary greatly among milk bottles.

Pasteurization   Pasteurization is a process that consists of the application of heat for a given holding time. Pasteurization is always defined by both temperature and time, and if temperature is increased the time requirement is decreased. The Food and Drug Administration (FDA) has provided several time-temperatures combinations,[5] which are designed to ensure a 5 log reduction of all vegetative bacteria in milk (Table 8.1). The objective of pasteurization is to extend shelf life and to ensure safety by killing most spoilage bacteria and all pathogens. Although successful at eliminating most vegetative cells, pasteurization is not capable of reducing viable spores, which is why pasteurized milk will still spoil within 20–30 days.

Two pasteurization techniques are commonly utilized. Batch pasteurization or low temperature long time (LTLT) utilizes a holding time of 30 min

TABLE 8.1
Pasteurization Temperature and Time Examples

Temperature (°F)	Temperature (°C)	Nomenclature	Time in Seconds
145	63	LTLT	1,800 (30 min)
161	72	HTST	15
191	89		1
194	90		0.5
201	94	Higher heat Shorter time	0.1
204	96		0.05
212	100		0.01

Source.   As defined by the Food and Drug Administration, 7 CFR Ch1 58.101.

minimum at a temperature of at least 145°F. Continuous pasteurization or high temperature short time (HTST) utilizes a holding time of 15 seconds minimum at a temperature of at least 161°F. LTLT involves filling a single vat with milk, heating it slowly to pasteurization temperature, holding it at the appropriate temperature for the required time, and then cooling using cold water in the jacketed vat. The vat is continuously stirred to ensure that every particle of milk is held at the required temperature. A headspace heater is utilized to ensure that all particles on the surface of the milk, such as in possible surface foam, are properly heat treated as well. Besides efficient stirring and headspace heating, the vat pasteurizer is equipped with safety features such as an indicator thermometer and leak detection valve (Figure 8.2). An external temperature recorder provides a permanent record and a legal

FIGURE 8.2
Batch (Vat) Pasteurizer with Air Space Heater, Indicator Thermometer, Leak Detection Valve, and External Temperature Recorder

LTLT Vat Pasteurizer

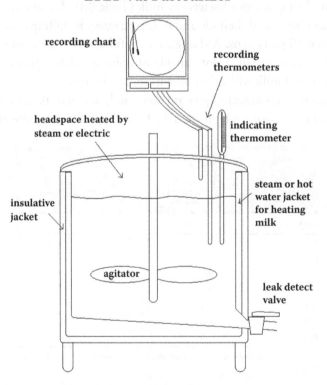

document of the pasteurization process. LTLT is both time-consuming and energy-intensive.

In contrast, HTST is quick and energy-efficient because the hot pasteurized milk heats the cold incoming raw milk, separated by stainless steel plates. Although HTST imposes more sheer force on the milk due to several pumps, heat exchangers, and high flow rates, the product quality generated by HTST pasteurization is widely accepted by U.S. consumers. The disadvantages of a HTST system are higher installation cost and the more complex process and equipment. Only licensed operators can legally pasteurize milk. Significant practical training is required prior to passing the HTST pasteurizer test, while a LTLT pasteurizer license is less difficult to obtain. Therefore most specialty dairy processors choose to utilize LTLT systems. Only when production is increased to the point of LTLT being impractical will the smaller companies switch to HTST systems.

Cheese

Cheese Types Specialty cheese types are many and varied, making the definition of cheese types challenging. Federal regulations largely lump cheeses into two general varieties, hard and soft. The soft cheese category includes soft ripened, soft, and semisoft cheeses, while the hard designation includes only very hard cheeses.[6] For labeling regulation there is also a limited number of cheese types found in the Code of Federal Regulations (CFR) Title 21:133.[1] Each cheese defined in the CFR includes a description of the finished product (fat content, moisture, ripening time), production techniques, required ingredients, optional ingredients, and labeling regulations. This list is incomplete and outdated. The original intention of this system was to protect consumers from imitation products; however, it now does relatively little other than define terms for use on product display panels. A good online version of the CFR is available at http://ecfr. gpoaccess.gov.

Specialty cheese producers define cheese types primarily based on how the product is produced in the vat and during aging. This works with the types of cheeses that have a well-defined production process, such as cheddar or gouda, but not for less common cheese types. The American Cheese Society breaks cheeses into 22 categories for their annual competition. After removing all categories that represent specific milk types and flavorings, the list looks like this: fresh unripened cheese, soft ripened cheese, American originals, American made/international style, cheddars, blue mold-ripened cheeses, Hispanic and Portuguese style cheeses, Italian type cheese, feta cheeses, marinated cheeses, and washed rind cheeses.[7]

Cheese Processing   It is fascinating that a bland-tasting product such as milk can be transformed into multitudes of products with bold characteristic flavors. Yet salt is the only added flavoring. Cheese flavors originate with the milk's proteins and lipids, coupled with the bacterial cultures used in the fermentation process. The flavors are "liberated" through fermentation and aging as the proteins and lipids are metabolized. Hundreds of different cheeses are made by basically the same process using the same four basic ingredients: milk, bacteria, rennet, and salt. Minor variations in the cheese making process and the ingredients account for major variations in finished products. Typical processes for cheddar, gouda, and camembert are summarized in Figure 8.3.

Cheese Milk Pretreatment   Few specialty cheese makers have centrifuges, and therefore they don't have the capability of standardizing milk to a constant protein/fat ratio, leading to inconsistent milk and cheese composition. In addition, cheese milk is seldom homogenized because it makes fat and moisture control in the cheese more difficult. Cheese milk may be raw, pasteurized, or thermized (heat treated). Thermizing of milk is a heat treatment that is not sufficiently intense to be within the pasteurization parameters defined by the FDA. This treatment is enough to kill most potential pathogens but leaves many native enzymes active, thus preserving flavor development during aging. By law, non-heat-treated and thermized milks are both considered raw. This means that the cheeses must age for a minimum of 60 days at a temperature above 35°F prior to commercial sale.

Milk Ripening   During this step in cheese production, milk is heated to a temperature that is conducive for the growth of lactic acid bacteria (LAB). This temperature ranges from around 90°F for mesophilic starter cultures to 110°F for thermophilic starter cultures. Starter cultures are added to the warmed milk and allowed to grow for anywhere between 30 min (just enough to overcome the lag phase of bacterial growth) and a full 24 h. Specialty cheese makers commonly add freeze-dried commercial cultures (direct vat set (DVS)) directly to the cheese vat. Growing a mother or bulk culture is not recommended because of the risk of contamination and phage (bacterial virus) infection. In addition, complex cultures lose diversity if they are frequently regrown because the strongest bacteria strains will outcompete and dominate the mix.

FIGURE 8.3
Flowcharts Outlining Processing Steps Involved in the Production of Cheddar, Gouda, and Camembert

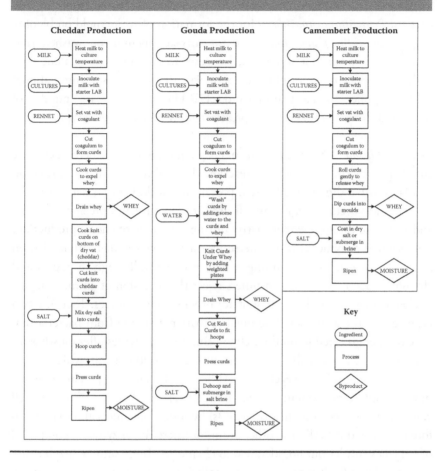

Different cheeses are made with cultures that optimize flavor and gas production at the unique processing and aging conditions. LAB convert lactose into lactic acid and are thereby responsible for acidifying milk for cheese production. The most common genera of LAB in cheese making are *Lactococcus*, *Streptococcus*, *Leuconostoc*, and *Lactobacillus*. It is common to add many species of each genus of bacteria in a complex starter culture to achieve desired fermentation rate and flavor production, along with lowering the risk for phage infection. Each specific species of LAB has the ability to ferment one or more substrates in milk, such as glucose, galactose, and citrate, while some produce CO_2, diacetyl, and other flavor compounds.

Adjunct cultures are not important during the initial milk ripening but are responsible for flavor development during cheese aging. Examples of adjunct cultures include *Propionibacteria* for formation of eyes and flavor in Swiss cheeses, *Penicillium roqueforti* for formation of the blue mold in blue cheeses, *Penicillium camemberti* for the formation of white mold on bloomy rind cheeses, *Brevibacterium linens* for surface ripening on washed rind cheeses, *Geotrichum* for soft ripened cheeses in combination or alone, and *Debaryomyces hansenii* for pH control on soft ripened cheeses. International culture suppliers such as Danisco and CHR Hansen offer a large number of single and mixed-strain cultures, preselected for specific cheese types.

Rennet Addition   "Setting the vat" or coagulation is a stage where an enzyme cuts kappa-casein and begins a reaction that culminates in the formation of a large three-dimensional casein protein network. Calf rennet, microbial rennet, and vegetable rennet are all used to denature and precipitate casein and form the casein polymer matrix during this stage of cheese production. The original enzyme used for cheese production was rennet, attained from an extract of calf stomach lining, which was naturally high in pepsin and chymosin enzymes. Some consumers reject the inclusion of animal enzyme in cheese, driving producers to look for other enzyme sources. There are two types of microbial rennet. Microbial rennets may come from bacteria that have been engineered to produce chymosin. These enzymes offer mostly similar characteristics to calf rennet without the inclusion of calf, which allows them to be considered vegetarian. Although the enzyme preparations are purified and do not contain the engineered bacteria, they are still considered genetically modified organisms (GMOs) by some consumers and may therefore be rejected as well. The other group of microbial enzymes is comprised of enzymes similar but not identical to chymosin. These similar enzymes are not considered GMOs and do not contain any enzymes previously found in calves. They are grown in a mold (*Rhizomucor* or *Aspergillus*) and then isolated for use in cheese. The final type of rennet is vegetable rennet, which is an isolate from various plants, with milk coagulation properties.

Regardless of the source, the activity of a rennet enzyme is dependent on pH, temperature, and time. Each enzyme reacts differently to changes in any of these milk characteristics, so experiments are needed to ensure proper application strength and set times when changing enzymes. Lactic acid accelerates the proteolysis of kappa-casein. This means that more acidic recipes generally utilize smaller doses of enzyme to achieve the same set rate. The specificity of each enzyme type is also different. Some fresh cheeses will react very well to an overactive enzyme, while such an enzyme could cause bitter off-flavors in an aged cheese. Therefore cheese makers have many viable rennet options for fresh cheeses, while calf rennet remains the best option for aged cheeses.

Milk that has been heat damaged during pasteurization does not set well because denatured whey proteins bind to kappa-casein and block rennet action. This problem can be partially overcome by increasing the amount of rennet and adding calcium chloride. Rennet is typically diluted with water (about 1 part rennet to 40 parts water) immediately prior to use. It is important that the water is not chlorinated and not overly hard, as both decrease rennet activity. The diluted rennet is mixed into milk utilizing the least amount of time necessary to ensure proper mixing. Overmixing will destroy the milk's ability to form a firm coagulum.

Cutting the Curd   Once set, the vat resembles milk Jell-O® pudding. This coagulum is cut into cubes to allow removal of whey. Curd sizes vary immensely, from not cutting the coagulum at all for a soft cheese, to cutting the coagulum into curds the size of rice kernels for a hard cheese. Smaller curds increase the surface area-to-volume ratio of the curds, accelerating the removal of whey from the protein matrix and resulting in a drier final product. Optimally all curds will be cut to exactly the same size. Some soft cheeses proceed straight from the setting stage to draining and hooping the cheese. This immediate drainage of the cheese ensures the highest possible moisture content by keeping the greatest amount of protein matrix intact. The firmness of the coagulum at the time of cut is also important in determining final moisture content in the cheese. Cutting a firm coagulum causes more moisture retention, while cutting a soft coagulum causes moisture loss. The knives utilized to cut the curd must be sharp to avoid shattering and subsequent yield loss.

Cooking the Curd   Immediately after cutting the coagulum, curds are soft and easily shattered. Most recipes will call for a rest period (healing) to increase the strength of curds before cooking begins. Curds are cooked to promote whey expulsion. During the cooking step the temperature is increased slowly, as rapid heating can lead to case hardening and product defects. Hard cheeses are generally cooked to a higher temperature and for a longer time than soft cheeses. For example, gruyere curds are cooked in the vat to above 135°F for over 45 min. In contrast, camembert curds are not cooked at all to maximize moisture retention.

Whey Draining and Molding/Hooping   Upon completion of the cooking step, curds are separated from the whey. This is typically performed by opening a valve at the bottom of the vat and allowing the whey to run out through a screen, but can also be performed by removing whey from the top of the vat by buckets or siphons. Unpressed cheeses are drained just to facilitate access to the curds, which are then placed in a perforated mold without cheese cloth, and allowed to drain overnight. With some cheeses, such as gouda, the curds are allowed to knit on the bottom of the vat before whey is removed, then the knitted curds are cut and placed in a cheese mold. The transfer of curds from vat to molds/hoops is called hooping. While this is generally a simple process, several problems can occur. Soft cheeses are generally small, which means that there are many molds to fill, which can take a long time. Soft cheese curds are rapidly changing due to moisture loss and acidification. Thus there might be a significant difference between the first and last cheese. To overcome this problem, smaller cheese vats (such as French coagulation vats) should be utilized. The best solution is utilizing vats that tip and empty into block molds within seconds. Another problem is slow filling of large molds, which allow time for individual curds to cool down. This leads to poor fusion of curd particles during pressing and openings or slits in the aged cheese. Faster filling, keeping the molds warm, and even heating the cheese making room are all factors that control the problem.

Salting   There are three common methods for salt application. The first is mixing salt into curds prior to hooping and pressing. The correct quantity of salt is weighed out and added to the curds while still in the vat. The curds are stirred to evenly distribute the salt, placed into molds, and pressed overnight. Addition of dry salt to the rind of cheeses is the second common method for applying salt to cheese. In this method salt is sprinkled or rubbed into the surface of cheese. The osmotic pressure of the dry salt pulls some whey out of the cheese and dissolves the salt into a small layer of high-concentration salt water on the surface of the cheese. The salt then slowly migrates toward the center of the cheese. This process may need to be repeated many times during the life of a cheese before desired salt content is reached. It can be challenging to ensure that the correct quantity of salt is added using this method. The final method for applying salt to cheese is through the use of salt brine. All brines should be adjusted to match the pH and calcium concentration of cheese. Only then should it be saturated with salt. It is common to brine cheeses at around 50–55°F. This method allows for the most precise control over the salt quantity absorbed by each cheese but requires more maintenance than other methods of salting. Due to the ability of *Listeria* to live in high-salinity environments and grow at low temperatures, it is important to monitor the microbial quality of any brine. Brines can last for a long time but may need to be cleaned by pasteurization or filtration.

Flavor Addition   Flavors, herbs, and other inclusions are added directly to curds prior to or during hooping. This often results in a pleasant marbled appearance of the aged cheese. Flavor addition to cheese can pose a risk due to potential contamination. For example, fresh herbs may contain pathogens such as *Clostridium botulinum*, which don't grow in the aerobic environment on herbs. However, when placed within the anaerobic interior of an aged cheese, *Clostridium* may be able to grow and produce toxins. Therefore all additions should be carefully reviewed on a case-by-case basis.

Temperature of the cheese making room is an important factor that is often overlooked. The temperature should be warm and consistent throughout the year. Many cheeses ferment for many hours after hooping. For example, soft cheeses generally remain in the cheese molds overnight. The temperature of the room impacts this fermentation. Consistent cheese can only be obtained if the temperature is constant. Pressing works best when cheeses are warm and pliable, which emphasizes the need for warm and consistent temperatures.

Aging   Cheese aging is as much a part of cheese production as anything that happens in the vat. The exact same cheese can become an aromatic epoise style cheese or a fluffy camembert simply due to changes in the aging environment and salt application. For proper aging, the environment must be controlled within tight parameters. Temperature should not fluctuate more than 1°F, as those changes can alter the activity of enzymes and determine which microbes are growing. Humidity and airflow must be tightly controlled, the combination of which influences the rate of moisture loss from the cheese. Air exchanges should be considered to remove buildup gasses and deliver fresh oxygen to the cheese. Last but not least, sanitation of the aging environment is essential for product quality. Even if aging cheeses are not in direct contact with contaminants, they can still absorb aromas from the surroundings. Systems for controlling these characteristics in a cheese aging facility can be as complex as remote-controlled programmable air control systems, and as simple as a pan in the bottom of a refrigerator to increase humidity. Most starting cheese makers face significant challenges optimizing the aging process because much of the knowledge about design and operation of small affinage facilities is largely anecdotal. It has been shown that most LAB involved in the aging of cheeses are not related to those inoculated as starter cultures in the beginning of cheese making. Because of this, they are known as non-starter lactic acid bacteria (NSLAB). Some NSLAB are added as adjunct cultures, while others are naturally occurring in the milk or environment. These noninoculated bacteria may be accountable for the specific terroir of a given cave, exemplifying the importance of a living cave in opposition to a sterile cave.

Other Cultured Products

Butter Churning butter is a method of reversing the emulsion of cream from fat in water to water in fat. Butter, as regulated by the FDA, must contain at least 80% fat; however, it is not uncommon for specialty butter to be closer to 85% fat. Regardless of the type of butter produced, a process of tempering the cream after pasteurization helps to ensure consistent texture and quality throughout the year. The University of Guelph's Dairy Science and Technology Education Series website provides specific suggestions for tempering of cream.[8]

There are several common types of butter produced in the United States for consumer sales. Commodity type butters are sweet cream butter, meaning they are churned from fresh cream. In contrast, many specialty butter products are cultured. They are produced from cream fermented with mesophilic LAB, which produce lactic acid and flavor compounds in the butter. The flavor profile of cultured butter can be customized with the addition of different LAB and represents a small yet growing U.S. market. Flavored butter is also a specialty butter product. Similar to cheese production, the added flavorings and herbs must be clean and safe. Butter readily picks up flavor compounds from the environment; thus care is needed to avoid transmittal of flavors between different samples. Due to the high lipid content, lipid off-flavors are common. Especially, oxidation can be an issue following oxygen and light exposure during storage. Fortunately, butter can be stored frozen for extended periods of time without noticeable quality loss.

Yogurt Yogurt is produced by pasteurizing standardized milk (milk proteins and stabilizers added) above typical pasteurization temperature to denature whey proteins and build viscosity. The milk is cooled to approximately 110°F and inoculated with thermophilic LAB (*Streptococcus thermophilus* and *Lactobacillus bulgaricus*). It is becoming increasing popular to also add probiotic cultures such as *Lactobacillus acidophilus*, *Bifidobacteria*, and others. Probiotic bacteria survive the stomach acidity and can temporarily inhabit the colon, which has been shown to improve immune function along with other health benefits.[9] Yogurt milk can be fermented in vats and subsequently cooled and packed (stirred yogurt), or the inoculated milk can be packed immediately into cups, which are stored in warm rooms until the fermentation is complete (set yogurt). The advantage of set yogurt is that generally it is not necessary to add stabilizers to obtain a firm texture; hence the label will be cleaner. The disadvantage is that heating an entire incubation room requires more energy than just keeping a fermentation vat at the appropriate temperature. Greek yogurt is currently gaining in popularity. It is produced in a process somewhat similar to chevre, as coagulated milk is drained for extended periods of time in cheese cloth bags.

Sour Cream and Crème Fraîche   The traditional difference between the two products is that crème fraîche does not contain stabilizers, and therefore is better adapted to incorporation in warm sauces and other dishes. Lately all-natural sour cream, with limited or no stabilizers, is gaining in popularity; thus the difference between the two product categories is disappearing. In both cases, pasteurized cream is cultured using mesophilic LAB. As with stirred and set yogurt, the fermentation can be done either in a vat or in the container. The key to good quality sour cream is to start with high-quality cream that is neither oxidized nor rancid.

Kefir   Traditional kefir is a fermented slightly acidic, carbonated, alcoholic milk drink. Typically, fermentation takes place through the inoculation of milk with kefir grains, which are symbiotic yeast and bacterial colonies contained within a biofilm. Kefir is not defined in the CFR and can therefore be made using different methods. Because of inconsistent quality and questionable safety of traditional kefir, most kefirs on the market today are produced in much the same way as drinking yogurt (produced as yogurt but homogenized prior to packaging) and do not contain alcohol.

Buttermilk   Buttermilk commonly describes two products in the United States. The traditional form of buttermilk is the by-product of churning cream into butter. This can represent a whole range of flavor profiles based on the microbial flora and the type of butter that the buttermilk comes from. Also available are cultured buttermilks. Cultured milk is the most common buttermilk available to consumers. This type of buttermilk is produced simply by adding mesophilic LAB cultures to low-fat milk, and allowing the milk to sour and thicken prior to cooling and packaging.

Packaging Dairy Products

Milk packaging is as varied as the pasteurization that it undergoes, ranging from sterile plastic jugs and cartons to reusable glass jugs. As with all dairy packaging, reduction of light and air exposure play major roles in ensuring that shelf life is optimized. Light-induced oxidation can rapidly ruin milk that is displayed under fluorescent lights without adequate protection from its packaging. It is very likely that postpasteurization contamination plays a major role in reduced shelf life of fluid milk. Because of this, sanitation on the downstream side of the pasteurizer is even more important than on the raw side.

The primary spoilage organisms in fresh products such as chevre, cottage cheese, sour cream, and cream cheese are yeasts and molds. Reduction of yeast and mold counts is the best and cheapest way to combat spoilage due to these organisms; however, there are also methods to extend shelf life with packaging. Modified atmosphere packaging (MAP) is frequently used to flush oxygen out of the air space of fresh products. By eliminating oxygen and replacing it with a mixture of CO_2 and N_2, growth of molds and yeasts can be retarded by the anaerobic environment and the pH effect of CO_2.

Aged fermented milk products can be packaged for a shelf life in excess of fresh milk products. The packaging should be designed by selecting a film that represents a healthy amount of oxygen exchange to allow a product to live, but not spoil. Plastic films or coextrusions of paper, plastic, and foil can be designed to allow selective permeation of light, CO_2, oxygen, and nitrogen. Each product has specific requirements for maintaining quality and should be matched with the proper film to protect it during distribution and display. Light-induced oxidation, causing pink discoloration and off-flavors, is a problem in all dairy products and can only be eliminated by reducing light exposure of the product.

Producing High-Quality Dairy Products

It is not possible to produce a high-quality product from low-quality ingredients. Unfortunately with dairy, it is very easy to produce a low-quality product even when starting with high-quality ingredients. To maintain quality of dairy products a processor must control several environmental factors, including shear, time, temperature, light exposure, aeration, and sanitation. In short, *promptly handling milk in a clean, gentile, and cool environment* will help to ensure that high-quality milk is not damaged before it makes it to the table.

Raw Milk Quality

From animal to package there are ample opportunities for compromising milk quality. In general terms, good quality milk has a relatively bland flavor with a slightly sweet taste. Milk should be refreshing with no lingering aftertaste. Parameters that impact raw milk quality are animal health and cleanliness, milking parlor sanitation, equipment sanitation, air and water quality, storage temperature and time, and mechanical abuse. In a perfect world there would be no bacteria in the milk and it would never undergo shear or exposure to light. In reality, milk contains bacteria that originate from the udder, the milking system (milk contact surfaces), and the environment. The Pasteurized Milk Ordinance (PMO) outlines legal requirements for the quality of milk.[5] Legally, grade A raw milk must have a bacterial load of <100,000 cfu/ml per producer and a somatic cell count (SCC) below 750,000/ml for cow and sheep milk and below 1,500,000/ml for goat milk. These legal limits are less strict than industry guidelines. Specialty milk producers frequently produce milk with microbial counts under 1,000 cfu/ml and somatic cell counts under 200,000. While there are no defined requirements for coliform counts in raw milk, it is important to reduce all gas-forming organisms in raw milk products, as they can cause undesirable gas and flavor defects. The final legal requirement for high-quality raw milk is the absence of beta-lactam antibiotics. Legally, any detectable amount of antibiotics in milk is illegal. Obviously, all other chemical residues, such as cleaning and sanitizer chemicals, aflatoxins, and pesticides, should be excluded as well.

Milk Composition

Milk composition is also a quality parameter and varies greatly due to animal species, breed, state of lactation, season, and feed. These changes are sufficient to pose real challenges when processing dairy products. The differences are mostly absent in large-scale dairy processing because of management practices such as staggered lactation, commingling of milk, and year-round feeding of grains. Specialty dairy processors have fewer tools available. Thus they may need to modify processing parameters to produce consistent products. An example of this would be to lower the set temperature in cheese processing for high-protein milks in the fall. There is not a single analytical test available that properly predicts overall milk quality. Instead, a combination of tests is often utilized. *Standard Methods for the Examination of Dairy Products*[10] outlines most standard dairy analysis techniques for compositional and microbial content. Common causes for milk quality failures are outlined in Table 8.2.

TABLE 8.2
Raw Milk Quality Problems, Their Causes, and Strategies for Prevention

Milk Flaw Characteristic	Cause	Solution
Aroma of feed in milk	Aromatic feed consumed prior to milking	Withhold aromatic feed prior to milking
Elevated microbial contaminants, elevated SCC, reduced shelf life from rancidity and bitterness	Mastitis	Find and remove mastitis-infected cattle from the milk pool
Elevated microbial counts with aroma, flavor, and texture problems	Sanitation failures in milking, shipping, or storage vessels	Check all CIP/COP systems for proper operation
	Milk is spending too much time above refrigeration temperatures	Ensure that chillers are cooling milk quickly and keeping it cold
		Ensure that SSOPs and GMPs are followed at all times
Rancid aromas and clumps of separated fat on surface of milk	Pump cavitation or physical abuse of milk can break down fat micelles and increase lipase activity	Use gentile pump type or eliminate the cause of the pump cavitation
Bitter off-flavors and rancid/sour aroma	Growth of psychotropic bacteria in milk	Use milk as soon as possible after milking
Fruity aldehyde aromas	Aeration or light exposure causes oxidation	Protect milk from sun and fluorescent lights
		Fill tanks completely for storage and shipping

Transit and Storage Time from Milking to Processing

The PMO[5] stipulates that milk in a bulk tank must be cooled to below 45°F within 2 h of the end of milking. However, it is recommended that high-quality raw milk should be stored below 38°F. Acceptable storage and transport time is based largely on the initial quality and storage environment of the milk. Due to the variability of milking equipment, microbial loads, and storage equipment, it is important to understand the systems being utilized when determining acceptable storage time. Raw milk contains psychrotrophic bacteria that grow in refrigerated milk and produce off-flavors. Best practice is to use all milk immediately after milking; however, holding times up to 72 h after milking may be utilized if the milk is kept very cold and protected from shear, light, and contamination. Although storing milk is never recommended, it may be unavoidable. Dairy processors generally do not want to process 7 days a week. This practice is physically exhausting and time-consuming. Collecting milk for 2 days before processing tends to be a compromise that preserves milk quality while saving time and effort.

Cleaning and Sanitation

All cleaning requires heat, chemicals, and mechanical action to adequately remove soil from a surface. When cleaning by hand, warm water and adequate unsoiled soap must be used, along with scrubbers and brushes to supply the mechanical force. A wide range of color-coded brushes are available, allowing for color coding brushes for each process. Processors define which color of brushes should be used for each type of cleaning; whether food contact, floor contact, incidental food contact, or drains, each should have a designated brush to prevent cross-contamination. Much of the equipment used in milk processing can be cleaned with a system called clean in place (CIP). CIP systems work by recirculating hot water that contains cleaning chemicals through all pipes, dairy hoses, and equipment. A rapid flow rate is needed to ensure turbulent flow, which supplies the mechanical action necessary for soil removal. Storage tanks and silos cleaned with a CIP system use a spray ball to distribute the cleaning chemicals throughout the inside of the silo. Spray balls should be inspected quarterly to ensure proper operation. Typically, CIP cycles will incorporate both an acid and a caustic wash cycle. The acid wash cycle dissolves and removes mineral deposits such as calcium, while the caustic cycle saponifies and removes fats. Food contact surfaces in a dairy processing facility should be the last areas cleaned to reduce the chance of postcleaning contamination. Only after cleaning is complete can sanitation with sanitizers begin. It is essential to dilute sanitizers to the appropriate nonrinse strength. If too dilute, the sanitizer doesn't work. If too concentrated, the sanitizer can pose a health risk or may corrode the stainless steel. Tests (strips or titration) must be utilized to verify sanitizer strength. It is recommended to sanitize both after processing and prior to start-up, although only the latter is legally required. It is strongly recommended to work closely with a chemical supplier to establish all cleaning and sanitizing procedures.

It is not sufficient to clean equipment and food contact surfaces. A clean and sanitary environment is another important aspect of making safe and wholesome dairy products. Placing foam sprays or footbaths at the doors of production areas can help reduce the movement of contaminants between production facilities, and especially help to keep production areas cleaner in farmstead environments. Footbaths should be set up to automatically replenish themselves during production times, making them a passive form of sanitation. Even with this sanitation measure in place, it is important to change footwear and clothing before transitioning from farm environment to production environment. Hand washing sinks should be placed in an area that is conducive to frequent washing. Sanitizer should be placed next to the hand washing sink as the final stage in sanitation for personnel. It is important to emphasize that sanitizers do not represent an alternative to washing, and should never be mistakenly used in place of hand washing. When cleaning equipment, walls, and floors by hand, use of a foaming cleaner allows for delivery of chemical cleaners for longer contact time than liquid methods. This foam is delivered to a hose, which allows spraying everything in the environment. While foam reduces the work of scrubbing due to the longer chemical exposure, it is still important to supply mechanical cleaning. Foam generators can be sourced for wall mounting or mobility. Humans should be treated as the single largest microbial threat in a production facility. Aprons or smock covers and hair nets should be worn to reduce shedding of airborne microbial particles by personnel working around open vats. Fingernails should be short and clean before entry into a processing area.

Minimizing Risk Factors in Specialty Dairy Processing

Each day over 6 million gal of milk are produced, processed into dairy products, and consumed in the United States. Hardly anybody ever gets sick from consuming these products and the dairy industry has a stellar safety record. This wasn't always the case. In the early 1900s, one-quarter of all food-borne outbreaks were associated with consumption of contaminated milk and dairy products.[5,11] The introduction of pasteurization coupled with much improved cleaning and sanitation has dramatically lowered the risks associated with dairy products. Yet, some challenges remain associated with specialty products, such as fluid raw milk, homemade queso fresco, and some other specialty cheeses.

This book focuses on the technical aspects of production of specialty food products. Incidentally, there are no technical processes that can ensure the safety of raw fluid milk sold in commerce, and numerous outbreaks have been associated with consumption of raw milk. However, it is well beyond the scope of this chapter to enter into the debate concerning the safety of raw milk. The topic has been extensively reviewed by many other authors; thus we recommend readers consult sources such as the recently developed website Real Raw Milk Facts,[12] and other sources.[13]

When reviewing the history of recalls and outbreaks associated with dairy products, cheese is clearly the highest-risk product among the family of fermented dairy products. Crème fraîche, yogurts, kefir, and others are rarely associated with food-borne pathogens. This is primarily because of the low pH, but also because the production processes are simpler, and thus post-pasteurization contamination is less likely to occur. Nevertheless, processors must remain vigilant, as emerging pathogens have been found to survive in acidic environments below pH 4.6. Primarily, *E. coli* O157:H7 has been shown to survive for extended periods in yogurt.[14]

Specialty cheese making is an ancient craft, yet it is still in its infancy in the U.S. food processing sector. Many specialty cheese processors have started up within the past decade. This presents a unique challenge because these new production facilities do not have established food safety systems. Everything has to be developed from scratch and frequently by people who do not have extensive food safety knowledge. Combined with the reality that specialty cheeses are high-risk foods[15] if not produced correctly, it is perhaps no great surprise that this sector has experienced repeated product recalls. American cheese enthusiasts have developed and perfected attractive and flavorful cheeses. But in this focus on quality, it has often been forgotten that safety is an intrinsic component of quality. Consumers' first expectation of a great quality cheese is that it doesn't make them sick.

Far too often the discussion on safety of specialty cheeses is limited to the debate on whether to pasteurize or not to pasteurize. However, safety goes well beyond this topic. A safe cheese must be produced using high-quality milk, in a facility that is constructed for ease of cleaning, with good quality 3A certified stainless steel equipment, using a process that minimizes risks of postpasteurization contamination, aged in a controlled cheese cave facility, and cut and wrapped in a sanitized protected environment. In addition, it is essential to understand that all cheeses are not created equal. For example, the risks associated with high-moisture fresh cheeses such as queso fresco are much greater than for pressed, aged, cooked alpine style cheeses. Other cheese parameters that impact safety include raw milk quality, salt in moisture, rate and extent of acidification, lactic acid cultures, cook step (heating of curd and whey in cheese vat), and aging.

It has been argued that a focus on sanitation and safety is what led American food down the path to sterility and cheese in a can. The argument is based on the false premise that we must choose between safety and quality. A thorough examination of Europe's cheese sector demonstrates that it is possible to achieve both. Most European cheeses are made in high-quality sanitary processes with exceptional control of cross-contamination and pathogens. Gone are the days where European consumers accepted products containing food-borne pathogens. One common misconception is that most European cheeses are made from raw milk. That's no longer the case. Raw milk cheeses are extremely rare in Northern Europe, and even in France only 12% of cheeses are made from raw milk. And the primary raw milk cheese in France is comté, a cheese that is not only cooked to 135°F for 45 min in the vat, but also pressed and aged for many months—and thus one of the low-risk cheeses alluded to above. Many of America's greatest cheese makers have also demonstrated that quality cheese can be made with a focus on safety. Thus the discussion in America needs to move beyond the idea that quality and safety are mutually exclusive terms.

The common pathogens of concern in dairy products are outlined in Table 8.3. Readers may find it useful to review FDA recalls of dairy products, which are available on several sites, including http://fda.gov. The common approach to assuring safety of dairy products is based on two principles: pasteurizing the raw milk and preventing postpasteurization contamination. Cheeses can legally be made from raw milk but must be aged for a minimum of 60 days at a temperature above 35°F prior to sale. Pasteurization by itself does not ensure safety of a dairy product; in fact, some studies have demonstrated a similar or higher level of contamination in washed rind pasteurized milk cheeses than the raw milk counterparts.[16] Thus processors must consider all aspects of the process from milk production to the final cut and wrap process. This includes plant construction, equipment design, product and personnel flow, cleaning and sanitation, employee training, personal hygiene, product and environment testing, and process control systems.[17]

TABLE 8.3
Common Pathogens Associated with Dairy-Borne Recalls and Outbreaks

Pathogen	Dairy Product	Contamination Source
Campylobacter jejuni	Raw milk	Raw milk source
	Fresh cheese	Dairy farm environment
Entero-virulent *E. coli* (such as *E. coli* O157:H7)	Raw milk	Raw milk source
		Dairy farm environment
Salmonella species	Raw milk	Raw milk source
	Pasteurized milk (cross-contamination from raw milk)	Dairy farm environment
		Processing plant environment
	Fresh and aged cheese	
Listeria monocytogenes	Raw milk	Raw milk source
	Fresh and aged cheeses	Dairy farm environment
	Ice cream	Processing plant environment
Staphylococcus aureus	Raw milk	Raw milk source
	Aged raw milk cheese	Dairy farm environment

The best approach to producing safe dairy products is prevention of contamination. This is the principle behind Hazard Analysis and Critical Control Point (HACCP). The principle of HACCP is to outline every single step in the process of each product. Subsequently all risks associated with each step are outlined. These risks are generally categorized as physical (e.g., metal and glass fragments), biological (pathogens), and chemical (e.g., antibiotics, chemicals, and allergens). The HACCP plan then systematically outlines how each risk is controlled. This can be through a critical control point, such as antibiotic testing of the raw milk, or it can be a prerequisite program such as Good Manufacturing Practices (GMP) and Sanitation Standard Operating Procedures (SSOP). GMP is the minimum requirement for the production of safe and wholesome foods. An example would be the requirement that employees must wash and sanitize hands after entering the production area. SSOP outline specifically how to fulfill the GMP. For example, cleaning and sanitation procedures for each piece of equipment should be printed and available for employees cleaning the equipment. The procedures should also outline how cleanliness is verified, by whom, when, and what to do if the equipment is not clean. The development of a HACCP plan along with the associated prerequisites can be a daunting task. Fortunately there is help available. For example, the American Cheese Society has developed guides to help cheese makers develop HACCP plans. Other resources, such as university extension programs, also provide help.

Starting a Specialty Dairy Processing Company

As mentioned above, specialty dairy processing is a relatively young sector in the United States. As a consequence, the industry is currently experiencing certain growing pains. Entrepreneurs interested in starting up a dairy plant face challenges such as milk sourcing, obtaining reasonably priced equipment, complying with county, state, and federal regulations, dealing with regulators who are used to inspecting large-size dairy facilities, obtaining good advice from qualified consultants, getting valid economic data for developing business plans, and obtaining funding. Many of these challenges will be overcome within the next decade, but until then entrepreneurs will likely face frustrations as they start up their businesses.

Figure 8.4 contains a brief flowchart of most of the licenses and permits that entrepreneurs must obtain prior to starting processing, along with other parameters to consider, such as water sourcing and waste treatment. The flowchart was developed based on requirements in Oregon, and even within

FIGURE 8.4
Common Steps Involved in Starting a Specialty Dairy Processing Facility*

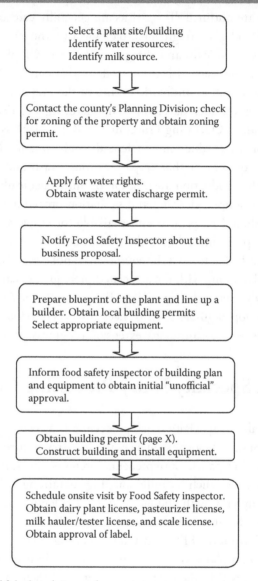

* County, state, and federal regulations apply.

this state local requirements such as zoning are different between counties. The figure and detailed information are available online.[18] Regulations vary somewhat among states, but mostly they follow the federal regulations administered by the FDA. Specifically, the FDA's PMO is commonly utilized.[5] Although the PMO is designed for grade A products (milk, yogurt, cottage cheese, etc.) sold in interstate commerce, many states have adopted the PMO for all dairy products produced within the state, including those not entering interstate commerce. So understanding and fulfilling the requirements of the PMO is generally a good place to start. Dairy licenses required for start-up may include dairy plant license, milk hauler and tester license, pasteurizer license, and scale license.

It is the experience of the authors that it is beneficial to work closely with local inspectors. Involving the inspector early in a start-up project helps prevent expensive problems and delays later on. For example, it is recommended to share blueprints of buildings and equipment before any investment is made.

Access to Milk

Obtaining milk is generally more difficult than expected. Most bovine dairy farmers are in an exclusive contract with a dairy cooperative or a private dairy processing company. The dairy company may be willing to sell the milk but at a premium price. The dairy entrepreneur must pick up the milk at the larger dairy plant with a licensed milk trailer and must have a milk hauler license. While it is difficult purchasing cows' milk, it is practically impossible to obtain goats' and sheep milk. Therefore many small-scale dairy processors have chosen to be farmstead operators, which means the milk is processed on the farm where it is produced. The advantages of being farmstead are easy access to fresh milk and full control over milk quality. The disadvantage is that extra precautions must be taken to ensure no cross-contamination between the processing and production sides of the operation. Clear rules must be established that nobody working with the animals enter the dairy processing facility unless they have washed and changed clothes and footwear. It may also be necessary to control the air quality in the processing rooms by ensuring a positive pressure of filtered air. This ensures each time a door is opened air will exit the processing room and airborne contaminants in the environment cannot enter.

Equipment

Small-size equipment is often nearly as expensive as larger equipment. There are fewer suppliers of small-size equipment, and because of the rapid growth of the sector there is limited availability of used equipment. Some people are choosing to work with importers of equipment from Europe where small-size equipment is abundant; however, the exchange rate and the need to convert equipment to U.S. standards (such as temperature monitoring on pasteurizers) likely result in the equipment being more expensive than domestic alternatives. No matter the source, it is highly recommended that blueprints of the equipment are evaluated by local regulators prior to purchasing to lower the risk of obtaining equipment that doesn't fulfill licensing requirements. A useful option is to hold off on paying the full price until the equipment has passed inspection. The U.S. dairy industry operates under FDA standards, and it would be expected that equipment licensed in one state could be licensed in another. Unfortunately this is not always the case. Local requirements or simply different interpretations of the same regulation have caused problems for several processors who have purchased used equipment from other states.

Waste Treatment

The liquid waste from dairy processors includes cleaning and sanitation chemicals, rinse water, and in the case of cheese production, whey. Chemicals have extreme pH values and must therefore be neutralized prior to release into municipal sewer systems or on-site disposal. pH neutralization may be as simple as mixing all liquid cleaning fluids in a tank, though some additional pH standardization could be required. Whey generally contains close to half the initial milk solids and specifically contains high concentrations of lactose, which contribute to the high biological oxygen demand (BOD). Only very small operators can afford to release high BOD loads into municipal systems. Large processors operate individual treatment systems. Artisan dairy processors are frequently left with neither option. A large range of creative wastewater solutions are observed, such as feeding whey to pigs, irrigating fields with wastewater, and spreading whey on grape vines for mold inhibition. All methods must be approved; releasing these liquids into surface streams is not a legal option. Frequently field application is not approved during winter months, which means some operators must have tanks to collect wastewater for up to 6 months per year.

Economics

The most frequently asked question during start-up is "How much will this cost?" Unfortunately very little information is available to forecast start-up and operating costs. As a consequence, start-up tends to be more expensive than expected. An unpleasant surprise for many people is that remodeling an existing building can be nearly as expensive as building a new one. Some estimates of building and equipment costs are outlined in Table 8.4, although the numbers are likely to fluctuate.

The break-even point is an indication of when a business turns profitable. It is recommended to break even within 3 years of start-up. When faced with large start-up costs many entrepreneurs choose to start out at a smaller scale. Unfortunately this can make expansion difficult, as an increased production of, say 30%, might require a 100% increase in investment as a new pasteurizer and other equipment may be needed. This business mistake can cause cheese operators to work extremely hard and still not have any chance of becoming profitable because not enough cheese can be produced in the facility. It is a better strategy to plan for growth by starting with overcapacity in both equipment and space and then plan for steady production growth. Some entrepreneurs plan for growth to reach 100% production capacity at the point when the equipment is fully depreciated, i.e., after 7 years of production.

A business model has been developed at Oregon State University to forecast start-up and operating costs and to determine the break-even point of potential artisan cheese operations. Sensitivity analysis reveals that the four factors that are especially important to business success are product price, product yield, milk price, and cheese maker wage. Entrepreneurs must have a firm understanding of what price their product can be sold at. This also means that the sales outlet decision is important. Common sales outlets include

TABLE 8.4
Major Start-Up Costs Associated with Specialty Cheese Processing

Construction of processing facility	$110/ft^2
Pumps and piping	$4,000
Raw milk and whey storage tanks	$9,000
Cheese aging room with shelving	$28,000
Vat pasteurizer and cheese vat	$37,000
Cleaning and sanitation equipment	$25,000
Drain table	$3,500
Cheese press	$6,000
Vacuum packer	$4,500
Cheese molds	$5,000
Van for cheese sales	$20,000

TABLE 8.5
Advantages and Disadvantages Associated with the Three Common Sales Outlets Available to Specialty Dairy Processors

Sales Outlet	Price to Cheese Maker	Advantages	Disadvantages
Direct sales (farmers' market, on-farm sale)	100%	Highest price Advertisement for cheese Direct contact with consumers	Labor intensive Seasonal Fuel cost Loss due to sampling
Wholesale (grocers and restaurants)	75%	All year Reach a larger, more traditional consumer group	Lower $/lb cheese Additional labor required to market and deliver directly to wholesale
Distribution	50%	All year Reach largest potential consumer base Least amount of sales labor required	Lowest $/lb cheese Must be able to supply the demand on a regular basis

farmers' markets, wholesale, and distribution. Each option has advantages and disadvantages, as outlined in Table 8.5.

The economic burden associated with starting up production of specialty dairy products along with the challenges associated with sales and distribution are frequently underestimated. Entrepreneurs start businesses because they enjoy producing specialty products such as cheese or ice cream. While they likely produce excellent products, it is the business side that ends up causing the greatest trials. A strategic plan that provides a realistic plan for growth is required for success.

Conclusion

This chapter summarizes some of the challenges associated with specialty dairy processing. The cautious reader could easily be discouraged from entering into this type of enterprise. Yet the benefits are abundant. As a farmstead dairy processor you would know your animals by name. You would not have a dreaded morning commute because your business is at your home. You would likely be living in a rural community surrounded with green pastures. The business would be a family business with your family functioning as a well-trained team with each member having unique responsibilities involving animal care, processing, and sales. Through contacts at farmers' markets you'd know the people buying and consuming your products. You would know that you contribute to sustainable rural communities. And most of all, you'd be producing a top-quality product, which should make you proud. So yes, the challenges are significant, but the rewards are plentiful.

References

1. Code of Federal Regulations. 2011. 21 CFR 110-133. http://ecfr.gpoaccess.gov (accessed April 15, 2011).
2. American Cheese Society. 2005. Cheese glossary. http://www.cheesesociety.org/i-heart-cheese/cheese-glossary (accessed April 11, 2011).
3. Becker, K.M., Parsons, R.L., Kolodinsky, J., and Matiru, G.N. 2007. A cost and returns evaluation of alternative dairy products to determine capital investment and operational feasibility of a small-scale dairy processing facility. *J. Dairy Sci.* 90:2506–16.
4. Colonna, A., Durham, C., and Meunier-Goddik, L. 2011. Factors impacting consumers' preferences for and purchasing decisions regarding pasteurized and raw milk specialty cheeses. *J. Dairy Sci.* 10:5217–5226.
5. Food and Drug Administration. 2009. Grade "A" pasteurized milk ordinance. http://www.cfsan.fda.gov/~ear/pmo03toc.html (accessed October 4, 2010).
6. Food and Drug Administration. 1998. Domestic and imported cheese and cheese products. Compliance and enforcement. http://www.fda.gov/Food/GuidanceComplianceRegulatoryInformation/ComplianceEnforcement/ucm071510.htm (accessed May 26, 2011).
7. American Cheese Society. 2011. American Cheese Society Judging and Competition Categories. http://www.cheesesociety.org/wpcontent/uploads/2011/02/2011 (accessed May 22, 2011).
8. Goff, D. 2011. Dairy Science and Technology Education Series. http://www.foodscience.uoguelph.ca/dairyedu/home.html (accessed June 15, 2011).
9. Sheil, B., Shanahan, F., and O'Mahony, L. 2007. Probiotic effects on inflammatory bowel disease. *J. Nutr.* 137(3):819–24.
10. Laird, D.T., Gambrel-Lenarz, S.A., Scher, F.M., Graham, T.E., and Reddy, R. 2004. Microbiological count methods. In *Standard methods for the examination of dairy products*, ed. H.M. Wehr and J. F. Frank, 153–87. 17th ed. Washington, DC: American Public Health Association.
11. Cornell University. 2007. Milk facts. Heat treatment and pasteurization of milk. http://www.milkfacts.info/Milk%20Processing/Heat%20Treatments%20and%20Pasteurization.htm#PastHist (accessed March 15, 2011).
12. American Veterinary Medical Association (AVMA) and International Association for Food Protection. 2010. Real raw milk facts. http://www.realrawmilkfacts.com (accessed April 4, 2011).
13. Griffiths, M.W. 2010. The microbiological safety of raw milk. In *Improving the safety and quality of milk*, ed. M.W. Griffiths, 27–54. Cambridge: Woodhead Publishing Ltd.
14. Dineen, S.S., Takeuchi K., Soudah, J.E., and Boor, K.J. 1998. Persistence of *Escherichia coli* O157:H7 in dairy fermentation systems. *J. Food Prot.* 61:1602–8.

15. Food and Drug Administration/Center for Food Safety and Applied Nutrition, USDA/Food Safety and Inspection Service, Centers for Disease Control and Prevention. 2003. Quantitative assessment of relative risk to public health from foodborne *Listeria monocytogenes* among selected categories of ready-to-eat foods. http://www.foodsafety.gov/~dms/lmr2-toc.html (accessed May 5, 2011).
16. Rudolf, M., and Scherer, S. 2001. High incidence of *Listeria monocytogenes* in European red smear cheese. *Int. J. Food Microbiol.* 61:91–98.
17. Meunier-Goddik, L., Bodyfelt, F.W., and Coelho Jr., J.R. 2009. Listeria prevention practices for small cheese operations. *Food Prot. Trends* 28:473–80.
18. Ravikumar, R., Durham, K., and Meunier-Goddik, L. 2009. Permits and licenses required for startup of artisan cheese plants in Oregon. OSU Extension Publication EM 8986-E. http://ir.library.oregonstate.edu/jspui/bitstream/1957/12569/1/EM8986.pdf (accessed February 5, 2011).

9

Juices and Functional Drinks

Olga I. Padilla-Zakour, Elizabeth K. Sullivan, and Randy W. Worobo

Cornell University
Geneva, New York

Ќ

Contents

Introduction

Juices and functional drinks (those that provide a health benefit beyond basic nutrition) are very popular products in the American market and include a variety of products, such as fruit and vegetable-based juices and drinks; teas; soy-, rice-, and almond-based drinks; yogurt drinks and smoothies; enhanced bottled waters; energy drinks; sports drinks; and so forth. On average, Americans consume 28 liters of single-strength (100% pure juice) fruit juice per year, with orange, apple, and grape juices being the category leaders (USDA, 2008). The juice and juice drink market reached $18.9 billion in 2010, but the consumption has remain unchanged in the United States since 2006, indicating that the market is in need of new alternatives to attract new consumers and increase overall consumption (Mintel, 2010b). The functional beverage market was valued at $9 billion in 2009, and it has been projected to reach $9.7 billion during the 2009–2014 period, reflecting a net decline after adjusting by inflation, mainly due to consumer economic concerns and lack of confidence on beverage health or performance claims (Mintel, 2010a).

The current dietary guidelines for Americans (USDA and U.S. Department of Health and Human Services, 2010) recommend the consumption of nutrient dense foods and beverages, and to choose foods that provide more potassium, dietary fiber, calcium, and vitamin D, as those are nutrients of concern in the American diet. The guidelines also recommend an increase in the consumption of fruits and vegetables and a decrease in the intake of added sugars, factors that favor 100% juices and beverages with a high fruit and vegetable content.

The main drivers for the juice market are health and wellness, and therefore consumers will choose beverages that offer high nutritional value and added health benefits, such as 100% juices fortified with vitamins, minerals, and functional natural compounds. Consumers also prefer all-natural beverages, without preservatives or artificial colors. The same trend has been observed in the European juice market. Interest in smoothies has increased in Europe as consumers are increasingly attracted to 100% not-from-concentrate juices due to their healthier image (Mintel, 2011). Children aged 6–11 are reported to be the biggest juice or juice drinks consumers in the United States, with 98% of them drinking any type of juice, although the frequency of drinking had remained unchanged from 2005 to 2009, perhaps due to the increased competition of other beverages such as teas, flavored waters, and sports drinks, and the relatively high cost of 100% juices. The high calorie content of juice and juice drinks was one of the main reasons for consumers choosing not to consume these beverages, followed by the presence of added sugars or high-fructose corn syrup (HFCS) or artificial additives (Mintel, 2010). For 2010, the major consumer packaged foods trends reported by food manufacturers were sodium/sugar/HFCS reduction at 29.9%, private label (retailer or supermarket own brand) at 28.3%, natural claims at 23.4%, and convenience at 14.6% (Toops and Fusaro, 2011).

Clearly, juices and functional beverages that can meet current consumer preferences can be successful in the marketplace as long as they offer nutritional value at a reasonable price.

Regulations Applicable to the Production of Juices and Functional Drinks

This chapter assumes processors are manufacturing products in the United States. Similar regulations for the food industry in general and the beverage industry in particular are in place in several countries. The following regulations are of particular relevance to juice and functional drinks producers in the United States.

Bioterrorism Act of 2002

Since December 12, 2003, nearly all food processing facilities must submit Form 3537 to the FDA to register their business under the Public Health Security and Bioterrorism Preparedness and Response Act of 2002. The FDA website has detailed information regarding who needs to register, as well as the online and paper versions of registration forms. As part of the act, processors are required to keep records designed to enhance the security of the U.S. food supply. The requirements of the rule state that manufacturers must keep records that identify the immediate previous source of all food and ingredients received by their facility and the immediate recipient of the products produced by the manufacturer and sold to a retailer or distributor. Products sold directly to an end user (consumer) are exempt from the requirement. Records, in either paper or electronic format, should be kept on premises or at a reasonably accessible location and must be kept by the manufacturer for 6 months to 2 years, depending on the shelf stability of the food product. The FDA may bring civil action or seek criminal prosecution for businesses that fail to establish and keep the required records.

Good Manufacturing Practices (Code of Federal Regulations Title 21, Part 110)

Compliance with good manufacturing practices and sanitation requirements is the foundation for safe food production. While the regulations cover good manufacturing practices and sanitation requirements in great detail, there are six basic areas the food processor must address:

1. Water used for processing and for sanitation of equipment, facilities, ingredients, and processing must be safe, potable, and of sanitary quality.
2. Food contact surfaces must be of appropriate material (stainless steel or food-grade plastic usually preferred), in good condition, and easy to clean. Food contact surfaces must be cleaned regularly, before processing, after interruptions, and sufficiently often to prevent contamination. Food handlers should wear gloves.
3. Cross-contamination must be prevented through sanitary practices, including proper storage of ingredients and finished product, personal hygiene of workers, and facility design (ventilation systems, cleanable floors and walls, etc.).
4. Protection of food, food packaging materials, and food contact surfaces from adulteration.
5. Control of employee health conditions and hygiene that could result in microbiological contamination of food, food packaging material, and food contact surfaces.
6. Exclusion of pests.

Acidified Foods Regulations (Code of Federal Regulations Title 21, Part 114)

These regulations address juices and functional drinks that are classified as acidified foods. Currently, juices and functional drinks may be classified as formulated acid foods if most of the ingredients are acidic in nature (mostly fruit ingredients), while they may be classified as acidified if more than 10% of their ingredients are low acid (pH above 4.6, such as vegetables, legumes, or nuts), and thus acid or acid ingredients are added in order to bring down the pH below 4.6 to control the growth of the most heat-resistant pathogen, the bacterium *Clostridium botulinum*. The acidified foods regulations require that shelf-stable products be reviewed, and the corresponding scheduled processes be issued, by a recognized process authority, a person or institution with expert knowledge and experience to make determinations about the safety of a food process and formulation. The regulations stipulate that prior to starting production, the processing establishment must register with the FDA, the schedule process for the acidified food must be filed with the FDA, and supervisory personnel need to be certified (successful completion of the Better Process Control School or equivalent). There are specific record-keeping requirements, acceptable methods for controlling and measuring pH, thermal processes to kill pathogens and methods to control spoilage microorganisms, and what to do in the event of a process deviation or problem with making the product.

Juice HACCP Rule (Code of Federal Regulations Title 21, Part 120)

Fruit juices made from acidic fruits were considered safe until a number of outbreaks in the 1990s were associated with the consumption of fresh apple and orange juices. The pathogenic microorganisms responsible for the illnesses were *Salmonella* species, *Escherichia coli* O157:H7, and *Cryptosporidium parvum*. As a result of these outbreaks, the FDA issued a final rule in 2001 entitled "Hazard Analysis and Critical Control Point (HACCP); Procedures for the Safe and Sanitary Processing and Importing of Juice."

The rule requires processors and importers of juices to establish a HACCP plan to minimize the risk of juice contamination with biological (microbial pathogens), chemical (toxins, pesticides, heavy metals), or physical (glass, hard plastics, or metal fragments) hazards. The HACCP plan must be established for each processing plant by a team of individuals familiar with the product and the processing conditions, including persons trained in juice HACCP. Any company producing 100% juices, nectars, or purees as final products, or to be used in the manufacture of other juices and beverages, must comply with the regulation. The rule specifies that the juice (just before bottling or after bottling) has to be treated with a process that achieves at least a 100,000-fold decrease in the number of pertinent pathogen(s) likely to be present in the juice. This requirement is known as the 5 log reduction performance standard. The pertinent pathogen is the most heat-resistant microorganism of public health concern that may occur in the juice based on historical accounts or scientific findings.

In the case of apple juice, for example, the processors must address critical control points to eliminate the risk of *E. coli* O157:H7 and *Cryptosporidium parvum*. For orange juice, the most likely risk comes from *Salmonella* species. Other juices might need to address pathogens such as *Listeria monocytogenes* for reconstituted juices or *Clostridium botulinum* if they are not acidic and their pH values are higher than 4.6, as in the case of vegetable juices. The most common approved processing methods to control microbiological hazards are thermal pasteurization and UV irradiation at 14 mJ/cm². For apple juice with pH values of 4.0 or less, the current recommendations for thermal treatments to achieve a 5 log reduction of *Cryptosporidium parvum* and *E. coli* O157:H7 are 160°F (71.1°C) for 6 seconds (recommended treatment conditions in New York), 165°F (73.9°C) for 2.8 seconds, 170°F (76.7°C) for 1.3 seconds, 175°F (79.4°C) for 0.6 seconds, or 180°F (82.2°C) for 0.3 seconds.

Citrus processors have the option of treating the surface of the fruit prior to juice extraction because it is unlikely that the pathogens will enter intact sound fruit under current industry processing practices. The surface decontamination treatment must be a 5 log reduction after the initial brush washing step and may be a cumulative 5 log reduction achieved with multiple steps that may include chemical disinfection, hot water dips, or specific juice extraction methods that reduce surface area contact during juicing (pinpoint extractors). Proper monitoring, verification, and validation procedures are necessary to ensure that the HACCP plan is effective.

In addition to microbial hazards, the rule also established the maximum level of patulin (a chemical hazard) allowed in a single strength juice at 50 ppb. Patulin is a mycotoxin produced by molds such as *Aspergillus clavatus*, *Aspergillus claviforme*, *Byssochlamys fulva*, *Penicillium patulum*, and *Penicillium expansum*, and if found in significant quantities in fruit juices, it might constitute a health risk due to chronic exposure. Patulin is most often associated with apple juice and is most commonly found with unsound, rotting fruit (blue rot or core rot) and must be prevented from entering the juice by sorting out decayed/moldy fruit or removing infected or damaged flesh prior to juice extraction.

There is one exemption in the HACCP rule for retail establishments, that is, establishments that provide juice directly to consumers. Even though these operations are not required to implement a HACCP plan and therefore might not pasteurize or UV treat the juice, they must label fresh juices with a warning statement that describes the risk of consuming untreated juices: "WARNING: This product has not been pasteurized and, therefore, may contain harmful bacteria that can cause serious illness in children, the elderly, and persons with weakened immune systems."

Strict sanitary conditions, good agricultural practices, adherence to current good manufacturing practices, and utilization of clean healthy fruit are necessary to minimize the presence of microbiological and chemical contaminants in fresh-pressed juices.

More information is available at the FDA website. For specific details of best practices for implementation of the juice HACCP rule, refer to the FDA's document *Guidance for Industry: Juice HACCP Hazards and Controls Guidance*, first edition, final guidance, released in 2004.

Product Labeling (Code of Federal Regulations Title 21, Part 101)

Food products, including juices and functional drinks, are required to be labeled in compliance with regulations in 21 CFR 101. The regulations define what information must be present on a label, how that information must be displayed, and the type sizes and prominence of information. At a minimum, the following apply:

- The principle display panel (PDP) must contain the product's statement of identity (usual name), and net quantity in ounces or gallons, and in milliliters or liters, depending on total volume.
- The information panel (the panel immediately to the right of the PDP) must contain the name and address of the manufacturer, packer, or distributor, the ingredient list in decreasing order of magnitude, and any required nutritional labeling or allergen declarations. If any of these items are already on the PDP, they do not need to be on the information panel also.
- If the beverage contains juice, the percentage of juice must be clearly displayed on top of the nutrition facts panel. To calculate the percentage juice when concentrate is an ingredient in a beverage or juice is made from diluting a concentrate to the original single-strength or unconcentrated soluble solids levels (measured as Brix values) found in the fruit, Table 9.1 is used as the main reference.

The FDA website has a food labeling guide available that will help explain the requirements, including those for nutrition labeling. Go to www.fda.gov and search for *Food Labeling Guide.*

Beverage Categories

For the purposes of this chapter, we are going to discuss four main categories of beverage products: juices/juice-based drinks, teas, formulated drinks, and carbonated drinks. These beverages are sold shelf stable, refrigerated, or frozen, depending on processing and packaging options applied during production.

Juice and Juice-Based Drinks

Juices and juice-based drinks can be further broken into fruit-based products and vegetable-based products. Any time a processor "juices" a fruit, vegetable, or even something like ginger or sugar cane, that processor is involved in making juice. As such, the processor and the juice product are subject to the juice HACCP rule (21 CFR 120). The rule applies to juices regardless of sugar or water added to result in the final product. This definition of *juice* is slightly different from the juice standard of identity used for labeling purposes; it is possible for a producer to make a product that cannot be labeled as "juice," but nevertheless is subject to the juice HACCP rule. If the juice is made from concentrate by addition of the appropriate amount of water, the final single-strength (unconcentrated) juice must meet the original level of soluble solids (measured as Brix values) found in the corresponding fruit. Table 9.1 lists the Brix level for most fruit juices as specified in CFR Title 21, Part 101.30. Juice-based drinks use juices as the main ingredient, but often include other ingredients, such as water, acids, sugars, herbs, spices, flavors, vitamins, minerals, stabilizers, and so forth. These types of drinks are not covered by the juice HACCP rule but will still need to follow good manufacturing practices to guarantee their quality and safety.

Fruit-based juices and fruit juice-based drinks are usually considered acid foods since most fruits are naturally acidic. Exceptions would be fruits such as cantaloupe, papaya, banana, or persimmon, which are not typically acidic. In these cases, acid may be added to give the final product a lower pH. If acid is added in order to lower the pH, the product may be classified as an acidified food. The advantage of acidic juices is that they only need to be processed at temperatures below 100°C to achieve shelf stability, as microorganisms die faster under acidic conditions.

Vegetable juice and vegetable-based juice drinks are generally either low-acid products or acidified products. This is because most vegetables (with the exception of tomato, which is actually a fruit) are low acid and have a high pH, above 4.6. The lack of acid in these juices means that microorganisms are

TABLE 9.1
Soluble Solids (Measured as Brix) of Single-Strength Juices*

Juice	100% Juice[a]
Acerola	6.0
Apple	11.5
Apricot	11.7
Banana	22.0
Blackberry	10.0
Blueberry	10.0
Boysenberry	10.0
Cantaloupe melon	9.6
Carambola	7.8
Carrot	8.0
Casaba melon	7.5
Cashew (caju)	12.0
Celery	3.1
Cherry, dark, sweet	20.0
Cherry, red, sour	14.0
Crabapple	15.4
Cranberry	7.5
Currant (black)	11.0
Currant (red)	10.5
Date	18.5
Dewberry	10.0
Elderberry	11.0
Fig	18.2
Gooseberry	8.3
Grape	16.0
Grapefruit	10.0[c]
Guanabana (soursop)	16.0
Guava	7.7
Honeydew melon	9.6
Kiwi	15.4
Lemon	4.5[b]
Lime	4.5[b]
Loganberry	10.5
Mango	13.0
Nectarine	11.8
Orange	11.8[c]
Papaya	11.5
Passion fruit	14.0
Peach	10.5
Pear	12.0

(Continued)

TABLE 9.1 (CONTINUED)
Soluble Solids (Measured as Brix) of Single-Strength Juices*

Juice	100% Juice[a]
Pineapple	12.8
Plum	14.3
Pomegranate	16.0
Prune	18.5
Quince	13.3
Raspberry (black)	11.1
Raspberry (red)	9.2
Rhubarb	5.7
Strawberry	8.0
Tangerine	11.8[c]
Tomato	5.0
Watermelon	7.8
Youngberry	10.0

* As specified in the U.S. Code of Federal Regulations, title 21, part 101.30.
[a] Indicates Brix value unless other value specified.
[b] Indicates anhydrous citric acid percent by weight.
[c] Brix values determined by refractometer for citrus juices may be corrected for citric acid.

much more resistant to heat, and therefore temperatures above 100°C, under pressurized conditions, are required to render the product shelf stable, typically in the 115–125°C range (Weddig et al., 2007). Since the majority of entrepreneurs cannot practically produce a shelf-stable, low-acid product, vegetable juices usually have acid added to them to lower the pH and make processing for shelf stability feasible. Juices like this are usually considered acidified foods. As an alternative, vegetable juices and vegetable-based juice drinks are sold as pasteurized refrigerated products with a very short shelf-life. The addition of approved preservatives may increase the refrigerated shelf-life.

Tea

Tea production usually begins with brewing the tea followed by the addition of sugar or sweetener, and the addition of acid in the form of fruit juice or organic acids such as citric. These beverages are almost universally classified as acidified products. The acid ingredients are often added by the processor for the dual purposes of flavor enhancement and to lower the final product pH to a point where making the product shelf stable is practical.

Teas to which no acid is added may still have a low pH if the tea for brewing contains naturally acidic ingredients, for instance, dried flowers such as rose hips or lavender, or dried fruits such as orange peel. Sometimes brewing tea with these ingredients is enough to lower the pH of the tea to a point that shelf stability of the final product is feasible. Alternatively, the pH may be low enough to result in a longer refrigerated shelf-life. Tea that is low acid (pH greater than 4.6) will need high-temperature processing under pressure, and thus cannot be made shelf stable with the equipment available to the majority of entrepreneurs and must be sold as a refrigerated product with a shelf-life of days/weeks or as a frozen product.

Formulated Drinks

Formulated drinks include yogurt-based beverages, soy, rice, or almond-based drinks, smoothies, sports drinks, energy drinks, and enriched or flavored water. These beverages are usually classified as low-acid or acidified foods. They are almost never formulated acid products. Formulated drinks are primarily water-based beverages to which dried ingredients, herbs, natural extracts, sweeteners, vitamins, minerals, flavoring, colors, or preservatives are added to achieve a beverage that purports to have life-style or health benefits. Generally, these beverages rely on claims regarding ingredients or health benefits in order to attract customers. Claims must be backed up with references to scientific research or nutrition labels. Claims must conform to language approved for use by the FDA (see the FDA *Food Labeling Guide*).

Carbonated Beverages

Carbonated beverages contain carbon dioxide to provide the fizz or effervescence associated with them, and are made with a syrup or juice base to which water and other ingredients are added. After all the ingredients are mixed, the beverage is pasteurized or sterile filtered followed by carbonation and bottling. Alternately, the formulated beverage is carbonated, bottled, and pasteurized afterward. Carbonated beverages are acidic and are typically shelf stable due to a combination of factors that involve pasteurization, carbonation, and the addition of preservatives. The final level of carbonation varies from below 2.0 (light carbonation) to 4.0 volumes of carbon dioxide, depending on the flavor and formulation. For example, colas, lemonades, ginger ales, tonic waters, and soda waters range from 3.0 to 4.0, while fruit-flavored and cream sodas fall in the 2.5 to 2.8 levels (Shachman, 2005).

Processing and Packaging of Juices and Formulated Drinks

The first question facing entrepreneurs when thinking about processing their product is whether to produce the product themselves, or whether to have the product produced for them by a contract packer (also called a co-packer). Things to consider when making this decision include the following:

- How much control does the entrepreneur want to have over the production of the product? If he or she wants to maintain a great deal of control, processing his or her own product may be the best option.
- Is there a facility available for the entrepreneur to produce in? Juices and formulated drinks must be produced in an approved/inspected facility—not a home kitchen. Sometimes these approved facilities are a challenge to find. Once found, the facility may or may not have the equipment needed to make the beverage.
- Is the processing facility adequate for the targeted production volume? A co-packer will have a minimum production requirement before taking on a new client, and this minimum might be higher than what the entrepreneur expected.
- Can the entrepreneur afford, and does he or she want, to be the responsible party for any processing-related regulatory compliance? If he or she manufactures the product, he or she will have to comply with all relevant regulations for such product.

Juice Processing

The production of juice at small scale starts with fresh or frozen fruits and vegetables, or with commercial concentrates. When fresh produce is used, it is important to select sound produce and to sort out decayed/diseased fruits and vegetables that will decrease the quality of the prepared juice. Figure 9.1 shows an example of berry juice processing to make a shelf-stable clear juice. Sorted produce is washed with chlorinated (25–50 ppm of free chlorine) potable water to reduce surface contamination. Cleaned and washed produce is then prepared for juicing based on the type of fruit or vegetable. The preparation may include pitting, grinding, heating to soften the tissue or extract desired components such as pigments, blanching to inactive undesirable enzymes that might cause oxidation and darkening, and enzymatic treatments to break up tissue components and decrease viscosity, thus increasing extraction efficiency. Enzyme suppliers

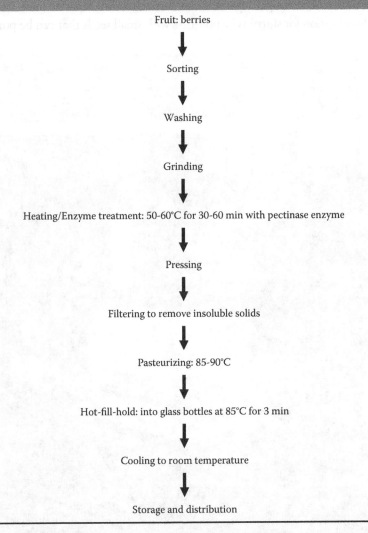

FIGURE 9.1
Diagram for Production of Clear, Shelf-Stable, Berry Juice Packaged in Glass Bottles

Fruit: berries

↓

Sorting

↓

Washing

↓

Grinding

↓

Heating/Enzyme treatment: 50-60°C for 30-60 min with pectinase enzyme

↓

Pressing

↓

Filtering to remove insoluble solids

↓

Pasteurizing: 85-90°C

↓

Hot-fill-hold: into glass bottles at 85°C for 3 min

↓

Cooling to room temperature

↓

Storage and distribution

are good sources of information regarding the type of enzymes and optimal conditions for their utilization. Grinding can be accomplished with hammer mills, grinding disk mills, graters, stemmer/crushers, or stoned fruit mills to reduce the particle size to 3–8 mm. Once prepared, the fruit or vegetable is pressed to separate the juice or fed into a pulper to produce a fine puree. There are several types of juice extractors available, including the basic rack and frame press, belt press, bladder press, basket press, screw press, horizontal piston press, and others. For small-scale operations, batch juice presses are utilized, shown in

Figures 9.2 and 9.3, with capacities of 200–300 liters/hour. For pulpy juices or puree production, pulpers or finishers are used, as they are capable of producing coarse and fine purees depending on the screen size selected (Figure 9.4). A decanter or centrifugal separator can also be used to separate juice, but this is only an option for slurry type products with small seeds that can be pumped

FIGURE 9.2
Hydraulic Horizontal Juice Press with Grater to Grind and Press Apples—Capacity of 200 Liter/hour*

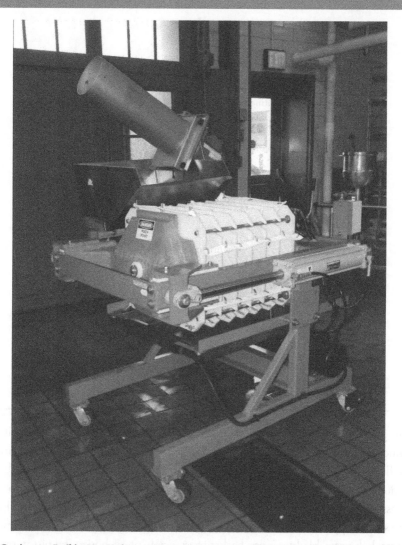

*Goodnature, Buffalo, New York.

FIGURE 9.3
Basket Hydraulic Press for Grape Juice Production—
Basket Capacity of 130 Liters*

* Mori, Italy.

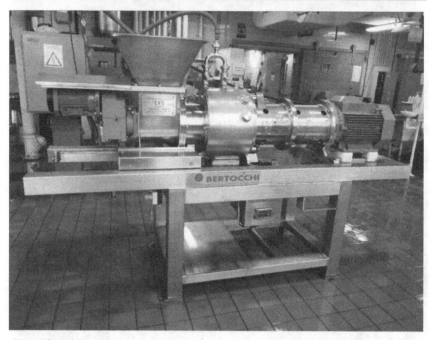

FIGURE 9.4
Extractor for the Production of Fruit and Vegetable Purees with Variable Coarseness Based on Screen Size—Capacity of 5,000 kg/hour*

*Bertocchi, Italy.

in a continuous manner, and normally requires a higher production volume to be economically feasible (Figure 9.5). The juice produced still has insoluble solids that will need to be removed if a clear juice is desired, which is commonly performed by filtration. Figure 9.6 shows a small plate filter typically used by small-scale processors. Cross-flow membrane filters are becoming more popular and are available in a variety of sizes and materials to fit many applications and production capacities. After filtration (if needed) the juice or puree can now be pasteurized to render the product safe, followed by cold or hot filling into selected containers, depending on the type of product being produced (shelf stable, refrigerated, or frozen) (McLellan and Padilla-Zakour, 2005).

Pasteurization and Stabilization

As mentioned earlier, juices and purees require a pasteurization step to destroy vegetative forms of pathogens of concern—the applicable pasteurization

FIGURE 9.5
Decanter for Separation of Juice from Fruit/
Vegetable Slurry—Capacity of 120–300 Liters/hour*

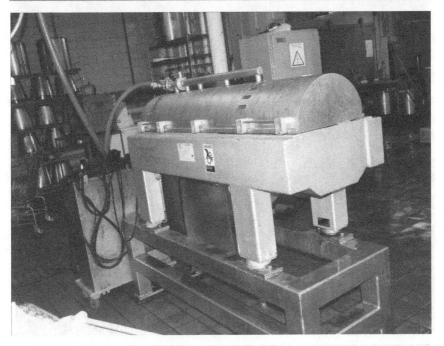

* Alfa Laval, Sweden.

regime will depend on the pH or degree of acidity of the product and the relevant microorganisms, but typical treatments include heating to 70–80°C for several seconds. This treatment will not render the products shelf stable, as spoilage microorganisms are more resistant than pathogens, although it is adequate for refrigerated or frozen products that are heat sensitive and would suffer from higher temperatures. To achieve shelf stability, the juice or beverage must receive enough heating to kill both pathogenic and spoilage microorganisms capable of growing under storage and distribution conditions throughout the expected product shelf-life. The two methods used in the production of shelf-stable acidic beverages are hot-fill-hold and in-bottle pasteurization (Weddig et al., 2007). For acidic products with a pH below 4.0, heating to 85–90°C and pouring the hot product into bottles, immediately capping and tilting or inverting the container to heat the cap, and maintaining the filled container hot for at least 3 minutes, will render the beverage shelf stable (Padilla-Zakour, 2009). It is critical to measure the hot-fill temperature in the filled

FIGURE 9.6
Small Plate Filter to Clarify Juices and
Beverages—Capacity of 50 Liters/hour

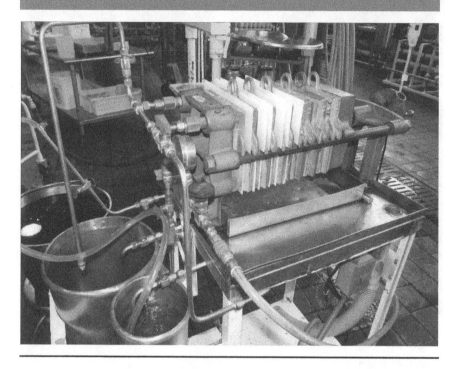

bottle (not just in the kettle or filling bowl) to ensure that the container and cap receive the proper heating to achieve the pasteurization regime designed for the product (see Figure 9.7). After the 3 min hot-hold period, the containers may be forced cooled with warm or cold water, or they can be slowly air cooled to reach ambient temperature. After proper inspection, containers can then be labeled and cased for distribution. The second option is to fill the bottles prior to heating the liquid and then subject the filled containers to in-bottle pasteurization, typically performed in a boiling water bath (100°C) or in a pasteurization tunnel for several minutes, as determined by the process authority. Carbonated beverages are pasteurized this way (Shachman, 2005).

Small-scale beverage processors must select equipment that fits their needs as well as their budget. Beverage production can start very simply, depending on the product, with a restaurant juicer and a stockpot on a stove. The product is juiced, heated to the appropriate temperature, and then hand-filled with a ladle into appropriate bottles, following the hot-fill-hold procedure. For larger production volume, the processing equipment needed, and

FIGURE 9.7
Talking the Hot Fill Temperature of a Beverage Right After Filling the Bottle

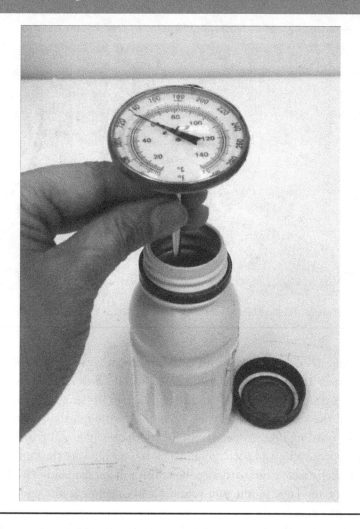

available, becomes more varied. One of the most common batch systems to heat and cook products is the use of steam stainless steel kettles, which are available in very small to very large sizes, and can be direct steam, electric, or gas fired to suit the needs of many processors. Figure 9.8 shows small 2.5-gallon (9.5-liter) and larger 20-gallon (76-liter) steam kettles for heating and cooking liquid foods, but they can also be used as water cookers for post-bottling pasteurization. Kettles can be purchased new or used from restaurant suppliers and equipment manufacturers or distributors. For continuous

FIGURE 9.8
Stainless Steel Kettles Heated by Direct Steam
Supply—Use for General Cooking and for
Heating Beverages to Hot-Fill Temperatures

Note.   Right side: 20 Gallon (76 liter) capacity; left side: 2.5 Gallon (9.5 liter) capacity.

processing, a variety of plate and tubular pasteurizers are available in the marketplace. The units are designed to meet specific flow rates (liters/minute) and pasteurization temperature and time based on the length of the holding tube, and the velocity at which the juice moves through the system, which feeds the bottling line. A special case of continuous thermal pasteurization is aseptic processing and packaging, which is used to make the familiar tetra brick or shelf-stable cardboard juice box. This process demands a high initial investment and large production volumes and therefore is not a good match for entrepreneurs.

In addition to thermal pasteurization, small-scale processors manufacturing refrigerated juice may utilize UV irradiation as the kill step for pathogens if proper validation procedures have been designed for those products. Juices with dark colors or high levels of solids and turbidity might not work well for UV units. For refrigerated apple juice, the FDA has approved the use of UV at 14 mJ/cm^2, a process that does not heat the juice or change the flavor of fresh-pressed juice (Tandon et al., 2003). The advantages of a UV system like the CiderSure (OESCO, Inc., Conway, Massachusetts) are simplicity of operation, minimum inputs (only electrical), and small footprint (see Figure 9.9).

FIGURE 9.9
UV Unit for Processing of Apple Juice to Achieve a 5-log Reduction of Pathogens—Processing Capacity of 100 to 500 Gallons Per Hour (380 to 1,900 Liters per Hour) Depending on the Model*

* Cider Sure, OESCO, Inc., Massachusetts.

Other nonthermal processing methods are available, such as high-pressure processing, dense phase carbon dioxide processing, and pulsed electric fields, but are not accessible to small-scale processors due to the high cost involved.

In order to increase the shelf-life of refrigerated pasteurized beverages to several weeks, chemical preservatives may be added, not to exceed the concentrations stipulated by the regulations or by good manufacturing practices. The most used preservatives are sodium benzoate and potassium sorbate, at levels not to exceed 0.1%, due to their low cost, lack of color, low toxicity, and efficacy against spoilage microorganisms under low (acidic) pH conditions (Davidson, 2001). A chemical sterilant that can be used during cold filling with specialized equipment to control the amount added to each container is dimethyl dicarbonate, an approved processing aid to decrease the microbial load of beverages. The compound acts very quickly and hydrolyzes rapidly in water to trace amounts of carbon dioxide and methanol; thus no labeling declaration is required. It has been reported to be effective against yeast and to a lesser extent other microorganisms at the maximum usage level of 250 ppm, and thus works well in beverages (Golden et al., 2005). It is possible to produce acidic shelf-stable beverages that rely on the combined effect of pasteurization and preservatives to achieve stability even though they are bottled at ambient temperatures. In this case, it is important to consult with a process authority to ensure both the safety and stability of the packaged drinks. Current work by the food industry on the development of natural antimicrobials with improved efficacy might result in more alternatives for the production of minimally processed natural drinks.

Formulated Drinks Processing

If the juice or formulated drink is produced from concentrate (juice that was evaporated to remove water) instead of fresh/frozen fruits and vegetables, then the process is much simpler, as can be seen in Figure 9.10. In this case, water is the main ingredient and the quality of the water becomes a key point to ensure that the final juice or beverage does not have objectionable attributes. The water may need to be filtered, de-chlorinated, or treated to remove chemical contaminants that might influence the quality of the juice or drink.

For formulated drinks, the process involves mixing all the components to meet the targeted specifications. The quality of the ingredients is of primary concern in order to produce consistent products that meet consumer expectations. It is very important to set up purchase specifications for each ingredient to ensure that low-quality materials are not used in the formulation.

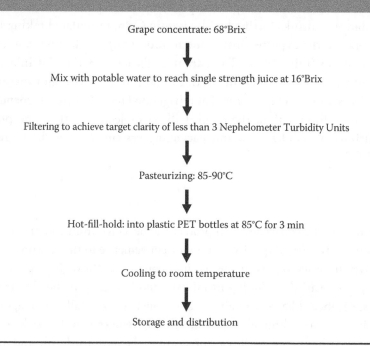

FIGURE 9.10
Diagram for Production of Clear, Shelf-Stable, Grape Juice Packaged in Plastic PET Bottles

Grape concentrate: 68°Brix

↓

Mix with potable water to reach single strength juice at 16°Brix

↓

Filtering to achieve target clarity of less than 3 Nephelometer Turbidity Units

↓

Pasteurizing: 85-90°C

↓

Hot-fill-hold: into plastic PET bottles at 85°C for 3 min

↓

Cooling to room temperature

↓

Storage and distribution

Ingredients that add functionality to the drinks need to be carefully selected, such as antioxidants, vitamins, minerals, extracts, concentrates, herbal components, fiber, specialty oils, high-intensity sweeteners, and stabilizers. It is also important to understand interactions that may occur when some ingredients are mixed together, as functionality may be lost or reduced by reactions leading to precipitate formation, oxidation, insolubility, or degradation. In addition, it might be necessary to balance the flavor of the final drink due to the strong flavor notes (bitter, metallic, earthy, or grassy, for example) contributed by the functional ingredients; thus careful testing of type and concentration of acids, sweeteners, and natural flavors will be crucial. Drinks that can claim low-calorie content (100 calories or less per serving), low or reduced sugar, and no high-fructose corn syrup are preferred by consumers, who are more likely to look for beverages with functional ingredients that include calcium (59% of consumers), antioxidants (51% of consumers), and vitamins (50% of consumers). Segments of the market that present opportunities for growth include beverages for Asians and Hispanics, as 65–67% of these consumers report using functional products (Mintel, 2010a).

Many functional ingredients are sensitive to heat, meaning that heating the drink even to pasteurization temperatures can negatively affect the expected functionality of the drink, and therefore the quality. Vitamin C, for example, degrades quickly when exposed to heat or UV light, and thus drinks fortified with vitamin C are formulated taking into consideration the expected losses due to processing, packaging, and storage conditions (Ellis, 1994). To guarantee efficacy, functional drinks may be dated with a shorter shelf-life than similar products without functional ingredients, or they may be stored at refrigerated temperatures. Depending on the final formulation, the drink will be subjected to the appropriate pasteurization, stabilization, and packaging system to meet the targeted shelf-life.

Packaging

The function of the package is to protect the beverage from contamination, to minimize the loss of quality, to provide convenience to the consumer, and to convey information about the product contained. Packaging options for beverages are varied, including metal cans and bottles, glass bottles, plastic bottles, cardboard-based containers, and pouches. Generally, beverages are filled into bottles, although for bulk sales, beverages may be packed into drums or canisters. For small-scale processing, the most used juice and drink bottles are those made from glass and plastic materials, as they can be purchased in smaller amounts and manually filled and capped. Refrigerated beverages are usually packed into food-grade plastic bottles. The plastic may only be able to withstand low temperatures, so the beverage must be below a certain temperature (specified by the bottle manufacturer) to ensure the bottle does not become distorted by exposure to hot liquids. Examples of bottles used for refrigerated beverages can be seen in Figure 9.11, made of high- and low-density polyethylene and PET. The lids on refrigerated bottles should have a tamper-evident seal. These lids may be the "push on" type of unthreaded lid, or they may be the threaded type of cap with a detachable ring often seen on soda bottles. Lids generally have a plastic insert or gasket to help reduce the possibility of the beverage leaking during distribution (Weddig et al., 2007).

Shelf-stable beverages are usually packed into either hot-packable PET, polypropylene, or glass bottles (Figures 9.12 and 9.13). These containers must be able to withstand the hot-fill temperatures required to make the beverage shelf stable. The lids must be capable of holding vacuum unless preservatives such as potassium sorbate and sodium benzoate are part of the formulation.

FIGURE 9.11
Popular Plastic Bottles (Polyethylene and PET) Used for Cold-Packed, Pasteurized, Refridgerated Drinks

Vacuum is very important in maintaining the hermetic seal, which keeps contaminants out of the product, and in preventing the growth of heat-resistant spoilage microorganisms like molds. Figure 9.14 shows a lug type metal lid with a plastisol gasket, and a continuous-thread plastic lid with gasket to make the seal. Caps with safety buttons are preferred by consumers, as the visual indication provides additional assurance that the product has not been opened (Weddig et al., 2007).

Quality Considerations

The initial quality of beverages is determined by the ingredients, and the processing and packaging conditions applied during production. The subsequent shelf-life of the packaged drink will depend on the storage and distribution conditions. The length of acceptable quality is set by the processor based on historical data, shelf-life testing, and market performance. Indicators of changes in quality include microbiological counts that represent the growth of spoilage microorganisms over time, undesirable chemical reactions

FIGURE 9.12
Polyethylene Terephthalate (PET) Plastic Bottles Used for Hot-Packed, Shelf Stable, Acidic Beverages

resulting in unacceptable quality, and undesirable changes in appearance, color, texture, or flavor.

Shelf-stable beverages are microbiologically stable, and therefore the shelf-life is dependent on physical, chemical, and sensorial changes that occur during storage, but most products are acceptable up to 9–24 months. Frozen products do not allow the growth of microorganisms and generally have a shelf-life of 12–18 months, based on color and textural changes. Perishable refrigerated beverages have a chilled shelf-life of days to months due to microbial growth and flavor changes (Ellis, 1994). When determining the shelf-life of beverages, sampling intervals at 20% of the expected life to collect six data points are recommended. Microbial counts of relevance to determine end of shelf-life in refrigerated products include total plate counts, yeast and mold counts, and lactic acid bacteria counts. Total plate counts of 1 million or higher per gram or milliliter typically will indicate the end of the shelf-life (Worobo and Splittstoesser, 2005).

FIGURE 9.13
Multiserve and Single Serve Glass Bottles Used to
Package Shelf Stable Juices and Drinks

Note.   Metal lids with plastisol gaskets are capable of holding vacuum to maintain a hermetic seal.

Physical and chemical measurements used frequently in the juice and beverage industries as quality and safety parameters include pH values, container vacuum, soluble solids content (Brix), total acidity (by titration), color (visual or with a colorimeter or spectrophotometer), viscosity, turbidity, and other measurements relevant to individual products, such as antioxidant content and individual vitamin or mineral concentrations. The pH or degree of acidity is measured with a pH meter (Figure 9.15) capable of reading accurately to two decimals and requires frequent calibrations with reference solutions or buffers to achieve accurate measurements. The container vacuum is a destructive measurement taken on a bottled beverage by piercing the cap with a vacuum gauge (see Figure 9.16). Vacuum values of 5 to 20 inches of mercury (17 to 68 kPa) normally indicate that the seal is intact (Weddig et al., 2007). The soluble solids content (% weight by weight) or Brix value is typically used as an indication of the sugar content of the juice or drink as sugars

FIGURE 9.14
Examples of Lids Capable of Holding a Vacuum

Note.   The gasket material on the underside of the caps.

are the predominant soluble solids in juices, although a correction might be needed if acids are present in high concentrations (important for citrus juices and drinks) (Shachman, 2005). The measurement is taken by a refractometer calibrated in the Brix scale, with models available as handheld and bench-top units (Figure 9.17).

FIGURE 9.15
Bench-Top pH Meter Used to Test the Degree of Acidity of a Beverage

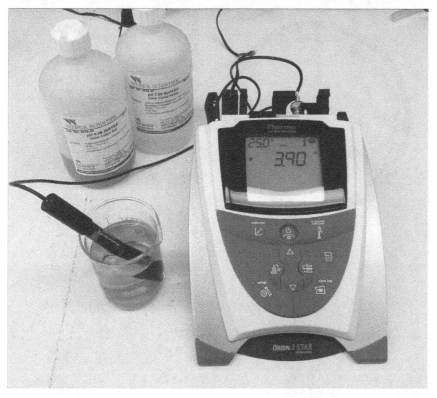

Note. The meter must be calibrated prior to use with two reference solutions (buffers).

Labeling

In addition to the five required pieces of information for the label mentioned above, beverage producers must consider two main factors: the name of their product and the claims about the product made on the label. The name of the product is strictly defined by the U.S. FDA. To be called a juice, a drink must be 100% juice. If the beverage is less than 100% juice, the beverage may not be called a juice, but may be called a beverage, drink, or cocktail (21 CFR 102.33(a)). Juices made from concentrate must be labeled as "from concentrate" or "reconstituted" (21 CFR 102.33(g)). Some exceptions are made for products like fruit punch and lemonade, although if individual juice components of the

FIGURE 9.16
Vacuum Gauge (Left); Reading the Vacuum of a
Shelf-Stable Bottled Juice (Right)

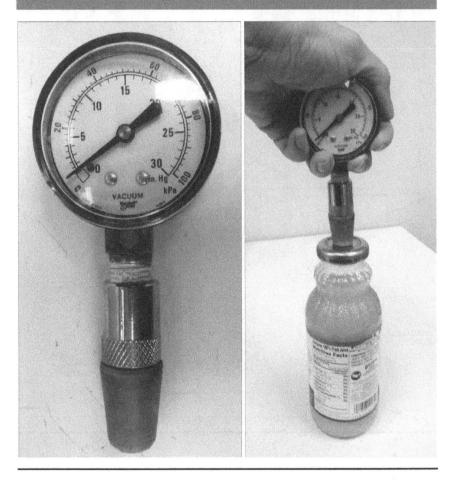

punch are from concentrate, and the juices are used in the product name, the juices in question should be labeled as such. The label for a mix of juices or a juice blend must name the juices in descending order by volume (FDA, 2009).

Some juices have definitions with which they must comply in order to be labeled "tomato juice," for example. These regulations are outlined in the following regulations: tomato juice, 21 CFR 156.145; beverages containing juices, 21 CFR 101.30; and definitions of lemon, lime, orange, pineapple, prune, and grapefruit juice and lemonade, 21 CFR 146. The definitions for juices stipulate things like color, soluble solids content measured as Brix, amount of added salt or sugar, etc. If the beverage does not meet the federal

FIGURE 9.17
Refractometers to Measure the Soluble Solids Content or Brix Levels of Juices and Drinks

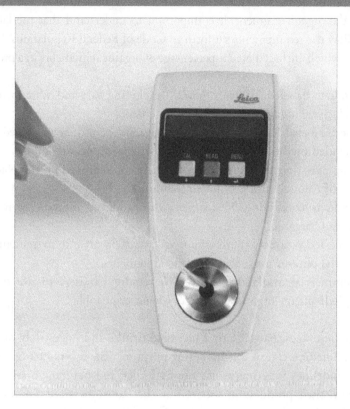

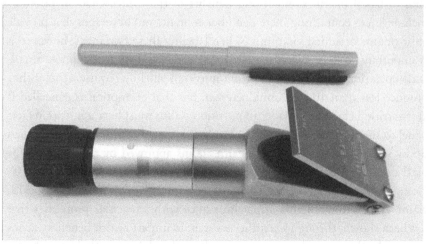

definition for a particular juice or beverage product, if applicable, the beverage cannot be label by the defined name. Alternatives such as "tomato cocktail" or "lemon drink" must be used.

Because many specialty juices and drinks are targeted to premium markets such as the organic segment, it is important to understand that the labeling must follow the requirements set forth in Code of Federal Regulations Title 7, Part 205, which are based on the percentage of organic ingredients in a product.

- Products labeled "100% organic" (excluding salt and water) contain exclusively organically produced ingredients.
- Foods labeled "100% organic" and "organic" cannot be produced using excluded methods (i.e., genetically engineered), sewage sludge, or ionizing radiation, or using processing aids not approved on the National List of Allowed and Prohibited Substances.
- Products containing 95% organically produced ingredients may be labeled "organic."
- Processed products that contain at least 70% organic ingredients can use the phrase "made with organic ingredients."
- Commercial availability provisions require handlers to use organic ingredients in "organic" products whenever possible.

Many beverages, particularly functional drinks and juices, rely on wellness and nutrition claims for marketing purposes. The recent changes in the dietary guidelines for Americans may lead juice and beverage manufacturers to label their products in such a way that consumers are able to make choices based on the content and benefits of the drink. Current offerings include juices containing increased fiber content, and beverages that provide antioxidants or added vitamins. When labeling these beverages, however, it is important to know what the label may claim and what it cannot. Small-scale operations may be exempt from nutrition labeling requirements if they produce less than 10,000 units per year, but that exemption is cancelled if any nutrition claims are made; all nutrition claims must have a nutrition facts panel to support the statement (FDA, 2009). Health claims must be backed up by scientific data, and only health claims approved by the FDA can be used on a label (21 CFR 101.9(k)(1), 101.14(c)–(d), and 21 CFR 101.70). Companies can petition the FDA to get approval to use a specific health claim. An example of this might be a juice with particular properties that has been shown, through scientific research, to impart health benefits to consumers. In this case, the company can base the claims on the research and petition the FDA to allow labeling that includes those claims.

References

Code of Federal Regulations. Current good manufacturing practice in manufacturing, packing, or holding human food. Title 21, Pt. 110. 2005ed.

Code of Federal Regulations. Acidified foods. Title 21, Pt. 114. 2005ed.

Code of Federal Regulations. Hazard analysis and critical control point (HACCP) systems. Title 21, Pt. 120. 2005ed.

Code of Federal Regulations. Food labeling. Title 21, Pt. 101. 2005ed.

Code of Federal Regulations. Common or usual name for nonstandardized foods. Title 21, Pt. 102. 2005ed.

Code of Federal Regulations. Canned fruit juices. Title 21, Pt. 146. 2005ed.

Code of Federal Regulations. Vegetable juices. Title 21, Pt. 156. 2005ed.

Code of Federal Regulations. National organic program. Title 7, Pt. 205. 2005ed.

Davidson, P.M. 2001. Chemical preservatives and natural antimicrobial compounds. In Doyle MP, Beuchat LR, Montville TJ, eds., *Food microbiology fundamentals and frontiers*. 2nd ed. Washington, DC: ASM Press, pp. 593–628.

Ellis, M.J. 1994. The methodology of shelf-life determination. In Man CMD, Jones AA, eds., *Shelf-life evaluation of foods*. London: Chapman and Hall, pp. 28–39.

FDA. 2009. *Food labeling guide*. http://www.fda.gov (accessed April 15, 2011).

Golden, D.A., Worobo, R.W., and Ough, C.S. 2005. Dimethyl dicarbonate and diethyl dicarbonate. In Davidson PM, Sofos JN, Branen AL, eds., *Antimicrobials in food*. 3rd ed. Boca Raton, FL: CRC Press, pp. 305–326.

McLellan, M.R., and Padilla-Zakour, O.I. 2005. Juice processing. In Barrett DM, Somogyi L, Ramaswamy H, eds., *Processing fruits science and technology*. 2nd ed. Boca Raton, FL: CRC Press, pp. 73–96.

Mintel. 2010a. Functional beverages—US (Internet). New York: Mintel International Group Ltd. http://www.mintel.com.

Mintel. 2010b. Juice and juice drinks—Europe (Internet). New York: Mintel International Group Ltd. http://www.mintel.com (accessed February 21, 2011).

Mintel. 2011. Juice and juice drinks: The consumer—US (Internet). New York: Mintel International Group Ltd. http://www.mintel.com (accessed February 21, 2011).

Padilla-Zakour, O. 2009. Good manufacturing practices. In Heredia N, Wesley I, Garcia S, eds., *Microbiologically safe foods*. Hoboken, NJ: Wiley, pp. 395–415.

Shachman, M. 2005. *The soft drinks companion: A technical handbook for the beverage industry*. Boca Raton, FL: CRC Press LLC.

Tandon, K., Worobo, R. W., Churey, J.J., and Padilla-Zakour, O.I. 2003. Storage quality of pasteurized and UV treated apple juice. *J Food Proc Preserv* 27(1):21–34.

Toops, D., and Fusaro, D. 2011. Food is changing in 2011. *Food Proc* 72(1):22–30.

USDA and U.S. Department of Health and Human Services. 2010. *Dietary guidelines for Americans 2010*. 7th ed. Washington, DC: U.S. Government Printing Office.

USDA. 2008. *Fruit and tree nuts situation and outlook yearbook 2008: FTS-2008*. Market and Trade Economics Division, Economic Research Service, USDA.

Weddig, L.M., Balestrini, C.G., and Shafer, B.D. 2007. *Canned foods: Principles of thermal process control, acidification and container closure evaluation*. 7th ed. Washington, DC: GMA Science and Education Foundation.

Worobo, R.W., and Splittstoesser, D.F. 2005. Microbiology of fruit products. In Barrett DM, Somogyi L, Ramaswamy H, eds., *Processing fruits science and technology*. 2nd ed. Boca Raton, FL: CRC Press, pp. 261–284.

10

Specialty Fruit and Vegetable Products

Yanyun Zhao

Oregon State University
Corvallis, Oregon

Ḱ

Contents

Introduction

A wide range of specialty fruit and vegetable products are available on the market and more are continuously introduced. According to the annual report *The State of the Specialty Food Industry 2011* by the National Association for Specialty Food Trade (NASFT),[1] from 2008 to 2011 the sale of specialty shelf-stable fruits and vegetables increased 11%, reaching a retail sale of $605 million in 2010; and specialty frozen fruits and vegetables increased 11.9%, reaching a retail sale of $256 million in 2011. Some of the unique characteristics of the specialty fruit and vegetable products may include the use of exotic fruit and vegetable varieties with distinguish flavors, such as hardy kiwi fruit and passion fruit; use of specialty ingredients in processed fruit and vegetable products, including all natural, gluten-, dairy-, and wheat-free; and application of novel and uncommon processing technologies that have minimal impact on fresh and delicious flavors and bioactive compounds of the products for health promotion. The specialty fruit and vegetable products include different categories of items from fresh and frozen to dried and shelf-stable products.

This chapter gives examples of different types of specialty fruit and vegetable products on the market, discusses the key technologies for manufacturing specialty fruit and vegetable products, and the controls for ensuring product quality and food safety, and provides some useful resources for obtaining additional information. Jelly types of products, such as jams, jellies, and preserves, were excluded from this chapter, as they are discussed in Chapter 6.

Ingredients Used in Specialty Fruit and Vegetable Products

Similar to other specialty food products, the uniqueness and specialty of fruits and vegetables are among the important considerations in developing specialty fruit and vegetable products. For example, more exotic fruit and vegetable varieties have been introduced into the U.S. market and processed into different forms of products, including frozen and dried products, vinegars, fruit sauce, and shelf-stable products, for providing consumers a wide range of choices. Examples of exotic fruit varieties may include Asian pears, dragon fruit, guava, passion fruit, pomegranate, mangosteen, quince, starfruit carambola, and shiro plum (Figure 10.1). In general, these exotic fruits have distinguished flavor, taste, color, appearance, texture, and high phytonutrients for providing additional health benefits to consumers. Also, more locally grown fruits and vegetables are promoted as specialty products, with a belief that they provide premier quality and are healthier than those from a distance.

In respect to the specialty of the vegetables, they may be classified into four categories: (1) ethnic vegetables that are grown quite commonly in other parts of the world, but are not yet known as a traditional staple crop in the United States; (2) baby or miniature vegetables that are baby and miniature versions of full-sized vegetables and not available or grown on a large scale for traditional market outlets; (3) off-season specialty items that are produced in high tunnels, or with other techniques to extend the traditional harvest season, i.e., early or late season extension; and (4) unusual or different varieties that are in demand by the marketplace and are mainly heirloom varieties, but may also be local cultural favorites.[2] The specialty ethnic vegetable market is growing throughout the United States, with Asian, Mexican, Puerto Rican, Indian, and African being the predominant market. Miniature vegetables are getting popular at farmers' markets and at fine restaurants as attractive garnishes for dishes, and they also have excellent direct market and roadside potential where they are seen as cute, novel, and unique, and may help as a good branding tool.[2]

FIGURE 10.1
Examples of Exotic Fruits that Are Commonly Seen at the U.S. Markets

| Passion fruit | Rambutan | Mangosteen | Dragon fruit |
| Start fruit | Jack fruit | Cherimoya | African cucumber |

In addition to the uniqueness and specialty of the fruits and vegetables, several other key aspects are also considered for the ingredients used in specialty fruit and vegetable products. These may include the following:

- All-natural or organic fruits and vegetables, as well as other functional ingredients used in the products
- No pesticide or synthetic fertilizers used in the production of the specialty fruit and vegetable crops
- Locally grown and premium quality of all ingredients
- Herbs and spices are used in processed fruit and vegetable products for flavors

The following sections discuss the key technology aspects for processing the mostly common specialty fruit and vegetable products (frozen, dehydrated, and shelf stable), give examples of specialty fruit and vegetable products on the market, and point out the possible controls for ensuring product quality and food safety.

Specialty Frozen Fruits and Vegetables

Except fresh, fruits and vegetables are most commonly processed into frozen form for long-term preservation or as pretreatment for further processing. As a method of long-term preservation, freezing is generally regarded as superior to canning and dehydration with respect to retention in sensory attributes and nutritive properties.[3] Freezing is generally considered to be the least destructive preservation technology for flavor, color, and phenolic compounds in fruits and vegetables and is recommended as a pretreatment for manufacturing juice or dehydrated product. However, the physical, sensory, and nutritional qualities of frozen fruits and vegetables are affected by the quality, variety, and maturity of the raw materials, freezing methods, packaging materials, and storage conditions.[4] In addition to the direct usage, frozen fruits are important ingredients in many other products, such as smoothies, pie fillers, muffins, pancakes, and yogurts, while frozen vegetables may be used for omelets, scrambled eggs, spaghetti sauce, meat loaf, stew, soup, and as steak toppers.

Specialty frozen fruits and vegetables are usually organic, premium, and ethnic brands. They are commonly freshly picked, precut for ready use, individually quick frozen (IQF), and packaged in clear polybags so that the quality and character of the product can be seen clearly by consumers, or with handy stand-up resealable bags and tubs for easy late usage. According to

Tanner,[1] the specialty brands of frozen fruits and vegetables may include Wyman, Willamette Valley Fruit, Remlinger, and Goya.

Methods and Equipment Used for Making Frozen Fruits and Vegetables

The methods and equipment used to make frozen fruits and vegetables directly impact the quality of the final frozen products. In general, quick freezing (>4°C temperature reduction per minute) is desirable for making high-quality frozen products, as it has minimal damage on the plant tissues, resulting in low drip loss and less change in the texture quality when the frozen products are thawed. This will eventually lead to less loss in sensory and nutritional quality of frozen products.

Individually quick frozen (IQF) processing is commonly used for processing specialty frozen fruits and vegetables because it helps retain the appearance, texture, nutrients, and other sensory qualities of the products due to its rapid freezing rate. IQF may be achieved by several different technologies and freezing equipment, such as air-blast freezer, fluidized bed freezer, and cryogenic freezer using liquid nitrogen or carbon dioxide as cooling media. Each of these systems can be different in respect to their freezing rate, equipment, and operation cost. The decision in selecting a best suitable system depends on the overall consideration. Detailed descriptions of each of these freezing systems are discussed by Barbosa-Cánovas et al. (2005)[5] and Zhao (2007).[4]

Pretreatment before the freezing process may be necessary for some fruits and vegetables to ensure product quality. For example, hot water or steam blanching is usually applied for vegetables, such as broccolis, corns, green peas, and green beans, to inactivate enzymes of peroxidase, lipoxygenase, and cysteine lyase, as these enzymes directly attribute to the development of off-flavor and off-odor during frozen storage. Inactivation of these enzymes before freezing would significantly control the quality deterioration during frozen storage. In addition, ascorbic acid may be applied during blanching for some light-color fruit, such as apples and peaches, to control color changes caused by the enzyme polyphenoxidase (PPO).

After the freezing process, storing the products at low temperature and preventing temperature fluctuation during storage are important for retaining high quality of frozen fruits and vegetables. When temperature fluctuates, ice recrystallization occurs and ice crystals grow bigger each time. The large ice crystal growth damages cells, causing mushiness and moisture to be pulled from the product, leading to high drip loss. Meanwhile, other quality losses are speeded up due to higher temperatures. The recommended frozen storage temperature is between −18 and −23°C.

Figure 10.2 is a general flow diagram illustrating the specific steps and controls for making high-quality individual frozen fruits and vegetables.

Packaging Considerations of Frozen Fruits and Vegetables

Since specialty frozen fruits and vegetables are mostly IQF products, the packaging considerations discussed here are only for IQF fruits and vegetables.

Proper packaging of frozen fruits and vegetables is critical for protecting the product from contamination and damage during storage and transportation, as well as preserving flavor, color, texture, and other important product qualities. Several factors should be considered in designing a suitable package

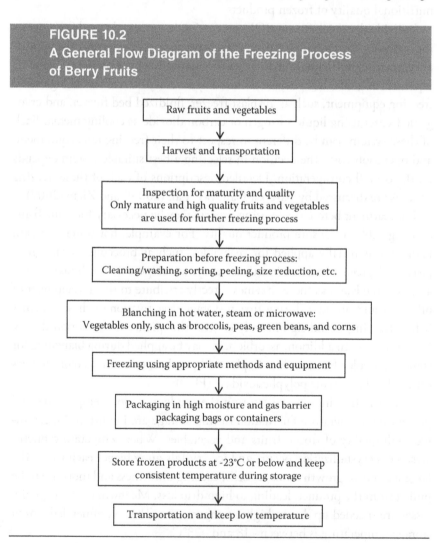

FIGURE 10.2
A General Flow Diagram of the Freezing Process of Berry Fruits

Raw fruits and vegetables

↓

Harvest and transportation

↓

Inspection for maturity and quality
Only mature and high quality fruits and vegetables
are used for further freezing process

↓

Preparation before freezing process:
Cleaning/washing, sorting, peeling, size reduction, etc.

↓

Blanching in hot water, steam or microwave:
Vegetables only, such as broccolis, peas, green beans, and corns

↓

Freezing using appropriate methods and equipment

↓

Packaging in high moisture and gas barrier
packaging bags or containers

↓

Store frozen products at -23°C or below and keep
consistent temperature during storage

↓

Transportation and keep low temperature

for frozen fruits and vegetables, including protection from external moisture, gases, and contamination, effective in terms of processing, handling, and cost, and attractive to the consumers.[5]

For retaining the highest quality of frozen fruits and vegetables, packaging materials should be moisture- and gas-proof to prevent evaporation of water and migration of oxygen from the environment. Fruits exposed to oxygen are susceptible to oxidative degradation, resulting in browning and reduced storage life. Therefore, packaging of frozen fruits is based on excluding air from the fruit tissues. Vacuum and oxygen-impermeable packaging materials are among the methods currently employed for packaging frozen fruits. Many moisture- and vapor-resistant wraps, such as heavyweight aluminum foil, plastic-coated freezer paper, and other plastic films, are effective at excluding oxygen. Plastic bags made especially for frozen foods are readily available. They can be sealed with twist-and-tie tops. Glass and rigid plastics are other examples of moisture- and gas-proof packaging materials.

The package should also be leakage-free and easy to reseal for providing convenience to consumers. Durability of the material is another important factor to consider, since the packaging material must not become brittle at low temperatures and crack.[5] Laminated plastic bags are the most commonly used package for frozen fruits and vegetables due to their flexibility during processing and handling. They can be used with or without outer cardboard cartons to protect against tearing. Glasses are not recommended because they can become brittle at frozen temperature.

Shape, size, and visibility of the package are also important factors in frozen specialty fruits and vegetables. Serving size may vary depending on the type of product, and the selection should be based on the amount of food determined for one meal. For shape of the container, freezer space must be considered since rigid containers with flat tops and bottoms stack well in the freezer, while round containers waste freezer space. Clear polymers or polymer-coated freezer paper with a clear window for viewing the products may be used.

Ensuring Quality and Food Safety of Frozen Fruits and Vegetables

Moisture migration is the principal physical change occurring in frozen foods, affecting the physical, chemical, and biochemical properties, including texture and palatability of the food.[6] Texture loss, freezer burn, and color and flavor deterioration caused by enzyme activity are the major quality losses in frozen fruits and vegetables.

Most fruits and vegetables have over a 90% total water weight. During the freezing process, when water in the cells freezes, an expansion occurs and ice crystals cause the cell walls to rupture. When thawed, the texture of frozen fruits and vegetables becomes much softer than that of nonfrozen ones. Consequently, drip loss is a result of thawing frozen products. Using a rapid freezing method for reducing the size of ice crystals and keeping temperature low and stable during frozen storage for preventing ice recrystallization help retain the texture quality and reduce drip loss.

Freezer burn, a condition defined as the glassy appearance in some frozen products produced by ice crystals evaporating on the surface area of a product, is one of the most common forms of quality degradation due to moisture migration in frozen foods. The grainy, brownish spots occurring on the product cause the tissue to become dry and tough and develop off-flavors. This quality defect can be prevented by using heavyweight, moisture-proof packaging during frozen storage.[6]

Several enzymes can cause the loss of color, nutrients, and flavor in frozen fruits and vegetables. These enzymes should be inactivated prior to freezing through blanching, a process that immerses fruits and vegetables in hot water or steam for a short time. In most cases, blanching is essential for producing quality frozen vegetables, since it also helps destroy microorganisms on the surface of the produce. However, in processing fruits, heat treatment may cause more degradation in quality. In this case, enzymes in frozen fruits can be controlled by using chemical compounds, such as ascorbic acid, in their pure form or in commercial mixtures with sugars for inhibition of enzymes in fruits.

Development of rancid oxidative flavors through contact of the frozen product with air is another potential chemical change that can take place in frozen products. This problem can be controlled by excluding oxygen through proper packaging, as mentioned earlier. It is also advisable to remove as much air as possible from the package to reduce the amount of air in contact with the product.[6]

In respect to food safety, it is important to understand that not all microorganisms are killed during the freezing process. In fact, some are killed, some are injured, and some may survive and can grow during thawing. The number of microorganisms surviving freezing depends on the number and type of the initial microorganisms, storage temperature, and other conditions. Therefore, the key controls for ensuring food safety of frozen fruits and vegetables are as follows:

- Following appropriate sanitation procedures during preparation and processing
- Prevention from cross-contamination from equipment and humans
- Keep low and stable temperature during storage

- Specific food safety program should be applied throughout the whole freezing process

Specialty Dehydrated Fruits and Vegetables

Dehydration is another traditional food processing and preservation method applied for fruits and vegetables. The principle of food dehydration is the removal of water from food materials using heat for water evaporation or sublimation so that the water activity of food is decreased to a level in which biological and chemical reactions are significantly delayed under normal storage conditions.

Specialty dehydrated fruit and vegetable products may include sun-dried (Figure 10.3a), juice, sugar- or salt-infused, dehydrated, and surface-coated items (Figure 10.3b), or freeze-dried products (Figure 10.3c) that well retain their flavor, color, and appearance. Dried fruits may be consumed directly as handheld snacks, or in many different applications, such as bakery items, confectionery products, cereals and granolas, food and nutrition bars, and trail mixes, while dried vegetables may be used as food ingredients in salads, vegetable trail mixes, soups, casseroles, and camping meals.

Technical Aspects of Making Dehydrated Fruits and Vegetables

Several textbooks have discussed different dehydration methods, equipment applied, and their impact on the final product quality in great detail.[7–10] This section briefly discusses the methods that may be applied for drying specialty fruits and vegetables.

Drying Methods Several different methods, including sun drying, hot-air drying, freeze drying, and infused drying, and various drying equipment, such as conventional hot oven, vacuum dryer, drum dryer, sprayer dryer, and

---

FIGURE 10.3
Dried Fruits and Vegetables Using Different Methods

Dried apricots	Infuse dried cranberries	Sun dried tomatoes	Freeze dried raspberries

freeze dryer, may be employed to remove water. The final moisture content and water activity, product quality, and processing cost can vary significantly depending on the specific method and equipment applied.

Sun drying is the oldest drying method, in which high-quality, matured fruits and vegetables are washed, sorted, cut, pretreated sometimes, and then placed on the trays to dry slowly under the sun. The process leaves fruits and vegetables with their natural color and concentrated with nutrients and flavor, and is low cost. The disadvantages may be low drying efficacy, little control on the weather, and potential attack by pests and microorganisms. Specific controls have been applied in modern sun drying for ensuring product consistence and food safety, such as applying a shield to prevent rain and nets for preventing the attack of birds and insects.

Hot-air drying is one of the most common drying methods and carried out in a mechanical device where food materials are static or in movement and hot air is conducted in different directions depending on the product nature. This may be the most used system for dehydration of fruits and vegetables because the process can be well controlled to permit a high quality of products with relatively low cost. Product feeding rate, air velocity, air humidity inside the heating apparatus, air recirculation conditions, and final moisture content in the product should all be controlled for obtaining desired product quality and drying efficacy. The hot-air drying may be done under vacuum so that the water can be evaporated at low temperature, thus helping retention of flavor, color, and nutrients of the dried products, but the cost of the system is higher than a conventional hot-air dryer. Various hot-air drying equipment is applied commercially, such as batch type for small-scale processes, continuous tunnel dryer for large-scale needs, drum dryer for puree type products, and spray drying for liquid products.[7-10]

One drying method that provides the best final product quality is freeze drying, a combined method that considers freezing first and then sublimation of water at very low pressure and temperature. In freeze drying, fruits and vegetables are first frozen at less than −35°C and then heated at low temperature to produce ice sublimation at a very low pressure, around a quarter-inch of mercury. This low pressure permits sublimation of ice at a very low temperature, a condition where fruit quality is well maintained. Due to the low temperature and sublimation of ice during the drying, freeze drying helps retain the nutrients, flavor, texture, color, and other sensory qualities. Unfortunately, the cost of freeze drying is higher than that of other dehydration methods, and thus the method is mainly used for products with high quality standards.

In the last decades or so, more infused dehydrated fruits have become available on the market.[11] In infused dehydration, fruit, whole or in pieces, is immersed in a sugar aqueous solution or fruit juice of high osmotic pressure. Due to the osmotic pressure difference, water flows out of the fruit into the sugar solution, and solute from the sugar solution transfers into the fruit. Depending on the type and concentration of sugar solutions, temperature, and immersion time, 20–50% of water may be removed during the osmotic process. After immersion, fruits are taken out, rinsed under water for moving surface sugar to prevent crystal formation on the surface, and then subject to hot air or other drying equipment to reach a targeted final moisture content. Oil, such as sunflower oil, is sometimes sprayed on the surface to prevent fruit from sticking together. Figure 10.4 illustrates the general procedures for

FIGURE 10.4
A General Flow Diagram of Infused Dehydration of Fruits and Vegetables

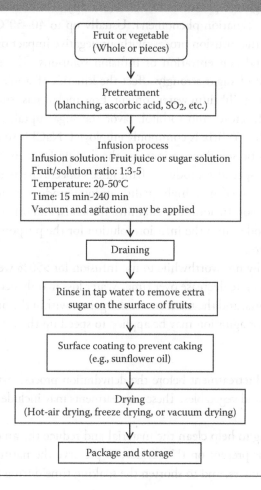

making infuse-dried fruits. This drying technology gives fruit soft texture as a result of the sugar impregnation into the pores of the fruit, provides a more desirable taste and flavor due to the addition of sugar and flavor from the fruit juice, and saves electrical energy by partially removing water through osmotic mass exchange. Moreover, this technology allows the impregnation of additional functional substances, such as nutrients, color, and flavor compounds into the infused product for meeting specific consumer and market interest.

Several process variables directly impact the processing efficacy and final quality of infuse-dried fruits and vegetables. These factors were discussed in great detail by Zhao and Xie (2004)[11] and are briefly stated below:

- Temperature. Rate of mass exchange increases with temperature, but above 45°C enzymatic browning and flavor deterioration begin to take place. High temperatures of >60°C modify tissue characteristics and favor impregnation phenomena. Usually up to 40–50°C is applied to speed up the infusion process without negative impact on fruit quality.
- Nature and concentration of infusion solutions. The type and concentration of sugar strongly affect the kinetics of water removal, solid gain, and equilibrium water content. Low molar mass sugars, including glucose, fructose, and sorbitol, favor the sugar uptake. A sugar solution of 60–75° Brix is commonly employed. NaCl is sometimes added into osmotic solutions to increase the driving force for water removal. For some specialty products, fruit juice concentrate is used to replace sugar for providing a high-quality, unique product with attractive flavor and color. In addition, vitamins, minerals, and other nutraceuticals may be added into the infusion solution for the purpose of nutritional fortification.
- It is usually not worthwhile to use infusion for >50% weight reduction as osmosis rate decrease over time. Water loss mainly occurs during the first 2 hours, and the maximum solid gain is within the first 30 minutes.
- Vacuum or agitation may be applied to speed up the infusion process.

Pretreatment   Pretreatment before the dehydration process may be necessary for some fruits and vegetables. These pretreatments may include the following:

- Blanching to help clean the material and reduce the amount of microorganisms present on the surface, to preserve the natural color in the dried products, and to shorten the soaking time during reconstitution.

Blanching can be done by hot water or steam. In hot water blanching, vegetables are immersed in boiling water in a wire basket for a short time period, usually 1–3 minutes, depending on the nature of the vegetables.[9] A 1% sodium bicarbonate and a 2% calcium chloride may be added into blanching water for protecting the bright green color of some vegetables and the texture of soft vegetables, respectively. However, such chemical preservatives are undesirable for specialty products.

• Use of chemical and natural preservatives, such as sulfur dioxide, ascorbic acid, citric acid, salt, and sugar, to preserve color and keep quality of final products. Many light-colored fruits, such as apples, darken rapidly when cut and exposed to air. If not pretreated, these fruits will continue to darken after they are dried. For long-term storage of dried fruits, sulfuring or using a sulfite dip are the best pretreatments. However, sulfites may cause asthmatic reactions in a small portion of the asthmatic population. Again, for specialty fruit and vegetable products, the use of chemical preservatives should be minimal. An ascorbic acid solution is a safe way to prevent fruit browning. However, its protection does not last as long as sulfite's. The mixture of ascorbic acid and sugar and honey solution may also be used to dip fruit before drying to prevent color darkening.

Quality Assurance and Food Safety

The basis of quality assurance is to prevent problems from arising, rather than trying to cure them afterward. Factors at every stage of the process, from raw materials to distribution of finished products, that could affect either product quality or safety should be identified, and specific procedures should be developed to monitor and control these factors so that they do not cause a problem.

For dehydrated fruits and vegetables, water activity of the finished product is a critical parameter that directly determines the shelf-life and microbial safety of the product. Water activity determines the available (free) water for microbial growth, enzymatic activity, and chemical reactions that change product quality and affect nutritional value, appearance, and acceptance of products. At a water activity below 0.85, most bacteria would not grow, but most molds and yeasts still grow at a water activity as low as 0.65. During the dehydration process, the water removal can cause several specific changes:

• Some microorganisms are destroyed by heat, and water activity is reduced by water removal. All these control the growth of microorganisms.

- Chemical changes such as browning caused by Maillard and carmelization reactions are a result of heating sugars. These reactions would affect the sensory quality of dried products. In addition, enzymes and proteins may be denaturized, and solutes are concentrated, which may cause the loss of nutrients, especially water-soluble vitamins.
- The removal of water causes food shrinkage; thus texture loss, unless freeze drying, is applied.

For quality assurance of dehydrated fruits and vegetables, the following specific controls should be considered:

- Moisture content and water activity should be controlled. Depending on the drying method applied, the final moisture content of the product varies. For example, hot-air-dried products usually have a moisture content of 10–12%, infused dried have 7–15%, and freeze drying can reduce moisture content to as low as 2–5%. Sensory qualities, especially color and texture, are also important quality indicators for dried fruits and vegetables.
- Packaging is critical. For preventing rehydration and oxidation during storage, high moisture, gas, and sometimes light barrier packages should be employed. Vacuum packages are commonly applied for excluding oxygen from the dried products.
- Although bacterial growth is usually not a concern for dehydrated fruits and vegetables due to their low water activity below 0.85, mold and yeast can still grow, especially mold that can grow at a water activity as low as 0.6.
- When the product is claimed all natural without any food preservative, it is important to pay attention to preventing mold growth on dried products, especially sugar-infused ones, as their water activity can be higher than 0.65. When exposed to oxygen, mold can grow.

The following microbial standards are usually applied for infuse-dried fruits by the industry: total plate count < 500 cfu/g, mold/yeast < 250 cfu/g, and coliform < 1 MPN/g. Depending on the final moisture content, package used, and storage conditions, the shelf-life of dried fruits and vegetables varies significantly. For example, 2 years for infused ones when vacuum-packaged and stored at cold dry conditions, and 20 years for freeze-dried ones with moisture content around 2% and vacuum packaged in cans or other high-moisture and gas barrier packages.

For ensuring food safety of dehydrated fruits and vegetables, the same controls as applied for other processed fruit and vegetable products should be applied, such as using high-quality mature fruits and vegetables, applying appropriate sanitation procedures, practicing human hygiene, and implementing specific food safety programs.

Packaging Consideration

Dried foods are susceptible to insect contamination and moisture reabsorption and must be properly packaged and stored immediately. For retaining high quality and long shelf-life of dehydrated fruits and vegetables, several key elements should be taken into considerations: (1) resistance of the packaging material to water and oxygen, (2) gas atmosphere inside package, (3) resealability of the package, and (4) storage temperature.

Dehydrated fruits and vegetables should be packed into clean, dry, insect-proof containers as tightly as possible without crushing. Dried fruits and vegetables can lose quality and spoil if moisture is reabsorbed and oxygen penetrates into the container during storage. Therefore, the packaging containers should be highly waterproof and have a gas barrier for protecting from the environmental moisture and oxygen. Glass jars, metal cans or boxes with tight-fitted lids, or moisture-vapor-resistant freezer cartons are good containers for storing dried foods. Glass containers are excellent for storage because any moisture that collects on the inside can be seen easily. Heavy-duty plastic bags are acceptable, but are not insect- and rodent-proof. Pouches using laminated materials, such as heavy foil and Mylar with zip-lock openings, have become more popular lately for packaging dried foods because of their light weight, resistance to water and gases, and resealability. Mylar is a trade name for a specific family of plastic sheet products made from the resin polyethylene terephthalate (PET). If sulfur has been used during fruit preparation and processing, such a product should not touch metal since sulfur fumes will react with the metal and cause color changes in the fruit. This way, fruit should be packaged in a plastic bag before storing it in a metal can.

Oxygen inside the package should be excluded since oxygen can accelerate the oxidation, thus shortening the shelf-life. Vacuum packaging is an option for dried fruits and vegetables packaged in plastic bags that are impermeable to gases and water. Airtight nitrogen-flushed pouches use a unique packaging process that removes the majority of the residual oxygen by immediately packing nitrogen into the Mylar pouch.

Note that each time a package is opened, the dried food is exposed to air and moisture that can lower the quality of the food and result in spoilage. Depending on the applications of dried products, they are packaged in a small-size package in which the contents are used immediately, or if a large amount of food is packaged and can't be used at once, good resealability of a package is essential.

The quality of dried fruits and vegetables is affected by heat; thus storage temperature determines the shelf-life of dried foods: the higher the temperature, the shorter the shelf-life. Most dried fruits can be stored for 1 year at 60°F (15.5°C), 6 months at 80°F (26.7°C). Vegetables have about half the shelf-life of fruit. Freeze-dried strawberries may be stored in a sealed #10 can with an included oxygen absorber packet for 10 to 15 years under ideal storage conditions (cool and dry).

Specialty Shelf-Stable Fruit and Vegetable Products

The retail sale of specialty shelf-stable fruits and vegetables was $605 million in 2010, an 11% increase from 2008, and shared 10.6% of the total specialty food market.[1] Shelf-stable products are those that have been processed in a way that they can be stored at room temperature while retaining quality and food safety before the containers are opened. Specialty shelf-stable fruit and vegetable products may include gourmet, ethnic, and natural canned fruits and vegetables; tomato sauce, pizza sauce, and tomato soup and puree that use all-natural ingredients and are gluten-free; and thermally processed, flavored fruit sauces. Wyman, Leroux Creek, Goya, and Dei Fratelli are the well-known brands of specialty shelf-stable fruit and vegetable products.[1]

Processing Technologies

Thermal processing is required by the FDA and USDA when making shelf-stable products that have a water activity above 0.85 because at this level of water activity, bacteria can grow, leading to food safety concerns. Depending on the pH value of the product, the level of required thermal processing is different. The term *pH* is a measure of acidity: the lower its value, the more acid the food. Acid may be naturally present in food, as in most fruits, or added, as in pickled food. Foods that are naturally low in acid (high in pH) can be acidified by adding lemon juice, citric acid, or vinegar.

Thermal Processing for Low-Acid Foods   Low-acid foods have pH values at or higher than 4.6, including red meats, seafood, poultry, milk, and all fresh vegetables except for most tomatoes. Based on FDA and USDA regulations, any food with a pH above 4.6 requires a canning process, called commercial sterilization, to kill bacteria spores of *Clostridium botulinum*. Basically, foods are packaged in tins cans or other heat- and pressure-resistant containers, hermetically sealed, and then subjected to pressured cooking at a given temperature and time combination in a retort. A retort is a closed, pressurized vessel for heating food sealed in containers.[12]

The key of thermal processing of low-acid food is to kill *C. botulinum*. Growth of the bacterium *C. botulinum* in canned food may cause botulism, a deadly form of food poisoning. These bacteria exist either as spores or as vegetative cells. The spores can survive harmlessly in soil and water for many years. When ideal conditions exist for growth, the spores produce vegetative cells that multiply rapidly and may produce a deadly toxin within 3 to 4 days of growth in an environment consisting of (1) moist, low-acid food, (2) a temperature between 40°F (4.4°C) and 120°F (48.8°C), and (3) less than 2% oxygen.

Botulinum spores are very hard to destroy at boiling water temperatures; the higher the canner temperature, the more easily they are destroyed. Therefore, all low-acid foods should be sterilized at temperatures of 240°F (115.6°C) to 250°F (121.1°C), attainable with pressure canners operated at 10 to 15 psig (pounds per square inch of pressure, as measured by gauge). At temperatures of 240 to 250°F, the time needed to destroy bacteria in low-acid canned food ranges from 20 to 100 minutes. The exact time depends on the kind of food being canned, the way it is packed into the containers, and the size of containers. Table 10.1 lists temperatures that apply for the thermal process of different types of foods, including canned ones.

Commercial canning kills all pathogenic and toxin-forming microorganisms that can grow under normal storage conditions. Canned foods may contain a small number of heat-resistant spores, but they will not grow under normal conditions, and the number of survivors is so low as to be insignificant. Canned foods usually have a shelf-life of 2 years.

Thermal Processing for Acid and Acidified Foods For products with a pH below 4.6, called acid foods, such as fruit products and tomato sauces with natural acid present, or acidified foods, defined by the FDA in 21 CFR 114.3(b) as "a low-acid food to which acid(s) or acid food(s) are added to produce a product that has a finished equilibrium pH of 4.6 or below and a water activity greater than 0.85," such as pickles and sauerkraut, a high-temperature process to destroy the spores of *C. botulinum* is not required because the spores of *C. botulinum* would not be able to reproduce at normal storage conditions or produce toxin at such a low pH level. The acid

TABLE 10.1
Temperatures Applied for the Thermal Process and Storage of Different Foods

Temperature(s)	Effect
240 to 250°F (115.6 to 121.1°C)	Canning temperatures for low-acid vegetables, meat, and poultry in a pressure canner
212°F (100°C)	Temperature water boils at sea level; canning temperature for acid fruits, tomatoes, pickles, and jellied products in a boiling water canner
180 to 250°F (82.2 to 121.1°C)	Canning temperatures are used to destroy most bacteria, yeasts, and molds in acid foods; time required to kill these decreases as temperatures increase
140 to 165°F (60 to 73.9°C)	Warming temperatures prevent growth, but may allow survival of some microorganisms
40 to 140°F (4.4 to 60°C)	**Danger zone:** Temperatures between 40 and 140°F allow rapid growth of bacteria, yeast, and molds
95°F (35°C)	Maximum storage temperature for canned foods
50 to 70°F (10 to 21.1°C)	Best storage temperatures for canned and dried foods

Source. Extracted from USDA, *Complete Guide to Home Canning*, Agriculture Information Bulletin No. 539, revised 2009, http://www.uga.edu/nchfp/how/general/food_pres_temps.html.

and acidified foods may be processed using a mild heat at the temperature of boiling water (100°C) or lower to destroy vegetative cells and some heat-sensitive spores, or in other words, to kill food-borne pathogens for ensuring food safety and destroy spoilage microorganisms for extending the shelf-life of a product. Due to mild, nonpressurized heat treatment, glass containers can be used. More detailed information on specialty acid and acidified foods can be found in Chapter 5.

Two common methods of heat treatment can be utilized for processing acid and acidified foods: (1) hot-fill-hold and (2) pasteurization. In the hot-fill-hold process, freshly prepared food is heated to boiling and simmered 2–5 minutes prior to filling, then promptly filled with hot product in the container (usually glass jars); the container is sealed and held for a given time period prior to cooling. The heat from the product will be enough to heat the container. Sometimes the containers will be inverted or tilted to provide heat to the container lid. The product may be heated in either a jacketed batch tank or by continuous processing using a high-temperature, short-time system.

For the pasteurization process, product, either cold or hot, is filled into a container, and the container is sealed. The container is then sent through a pasteurizer that uses flowing steam or hot water to heat the product and

the container together. The product is held in the pasteurizer at a specified temperature for a given time period to ensure destruction of the target
microorganisms.

Whether a product is processed using hot-fill-hold or a pasteurization
procedure, the juice, syrup, or water to be added to the product should
also be heated to boiling before adding it to the containers. This practice helps to remove air from food tissues, shrinks food, helps keep the
food from floating in the containers, increases vacuum in sealed jars, and
improves product shelf-life. Preshrinking food permits filling more food
into each jar.

Package of Shelf-Stable Specialty Fruit and Vegetable Products

For shelf-stable food, an airtight package (container) to produce and maintain hermetical sealing to protect against entry of microorganisms and
maintain commercial sterility of its contents during and after processing is
essential. The package must be sturdy and durable, and capable of remaining hermetically sealed under commercial operation and distribution conditions. Metal cans, glass jars, semirigid and rigid plastics, flexible pouches,
and paperboard may all be used. For specialty low-acid shelf-stable products,
metal cans are mostly used for sustaining the high heat and pressure, while
glass jars with unique and attractive shape that can clearly show the inside
content of the food are commonly used for acidified specialty shelf-stable
vegetable products, and semirigid plastic containers are used for fruit sauce
type of products.

Quality Assurance and Food Safety

Critical Controls   For ensuring food safety and obtaining high quality of
specialty shelf-stable fruit and vegetable products, critical controls should be
implemented, including the following:

- Initial product quality and cleanness
- pH control for acidified food (4.6 or below)
- Heat application, including temperature, time, and pressure, if it is for
  low-acid foods
- Vacuum
- Container sealing (seams for tin can and glass closure)
- Cooling
- Storage

Initial product quality should begin with high quality of fresh fruits and vegetables suitable for canning or pasteurization. Quality of fruits and vegetables varies among the varieties and changes depending on the stage of maturity. Freshness and wholesomeness of fruits and vegetables should be examined carefully before subjecting to process. Low-quality and contaminated fruits and vegetables should be discarded.

For making acidified food, pH control is critical for ensuring food safety. The acidification process should be monitored by taking pH measurements before and after equilibrium pH has been achieved. The critical factor is that the finished product equilibrium pH must be 4.6 or below. Finished product equilibrium pH represents the pH of the product (components included) in the final container after thermal processing, but not the raw product pH. The pH measurements must be recorded and the records should be reviewed at the appropriate time intervals.

Heat Application Depending on the pH value of the product, different levels of thermal process may be applied (Table 10.1). As stated above, for low-acid food with pH above 4.6, a canning process using high heat under pressure should be applied to destroy the spores of *C. botulinum*. A retort is used for this purpose. For products with a pH below 4.6, mild heat at boiling water temperature of 100°C or below at a given time period may be applied to kill food-borne pathogens and spoilage microorganisms for ensuring food safety and extending product shelf-life. Exact processing temperature and time and related operation procedures for a given product should be developed by the processing authority and should follow the instructions from the equipment manufacturer.

Vacuum Formation Vacuum formation inside the container is essential for ensuring food safety and quality. The purposes of creating vacuum inside the container include (1) minimizing strains on the seams due to expansion of air during the processing period, (2) removing oxygen, which accelerates corrosion in the can and causes oxidation of the food, and (3) reducing the destruction of oxygen-sensitive nutrients, such as vitamin C. Different methods may be employed for creating vacuum inside a container, including the following:

- Fill heated product into the container.
- Heat container and contents after filling.
- Evacuate headspace gas in a vacuum chamber.
- Inject superheated steam into headspace inside a container.

The most common and easy way is to leave a headspace inside the container and heat the products. When the product is cooled down, the vacuum is formed.

Container Sealing/Closure   Cans should be double-seamed as soon as the correct central temperature has been attained. Any delay between exhausting and seaming will lead to loss of vacuum and may lead to bacterial spoilage. Quality of the double seam for cans must be frequently checked. Avoid using cans or other containers with defective flanges, and jars with damaged lugs or threads. Periodic inspections must be conducted and records kept for ensuring that containers with defective seals are not produced.

Cooling   Thermally processed containers are usually cooled in water, and should be cooled to 37°C as soon as possible. This temperature allows water droplets on the can to evaporate before labeling and packing. Water of good sanitary quality must be used. Chlorination or other suitable sanitizers are employed to prevent contamination. Cooling can be done in retorts, cooling canals, spin coolers, rotary pressure coolers, or a combination of these. Cans processed in steam develop high internal pressure because of the expansion of foods, the expansion of air in cans, and an increase in vapor pressure of the water in the can. Glass, flexible, and semirigid packages and some larger metal containers almost always require cooling under pressure to prevent the seal from being damaged. However, too high a cooling system pressure or pressure maintained for too long of a time can cause container deformation.

Storage   Cans should be stored in cool and dry conditions. Maintenance of a constant temperature is desirable, as a rise in temperature may lead to condensation of moisture on the cans, with possible rusting. Cool conditions are required since storage at higher temperatures not only causes chemical and physical changes in the product, but also introduces a risk of thermophilic spoilage.

Procedures for Maintaining Color and Flavor in Shelf-Stable Fruit and Vegetable Products

The keys to maintaining good natural color and flavor in shelf-stable fruit and vegetables products are to[13]

- Remove oxygen from food tissues and jars
- Quickly destroy the enzymes in fruits and vegetables
- Obtain high vacuums inside the containers

For achieving the above goals, the following guidelines may be followed during processing and storage:

- Use only high-quality fruits and vegetables that are at the proper maturity and are free of diseases and bruises.
- Don't unnecessarily expose prepared foods to air, and process them as soon as possible. If fruits and vegetables are not processed immediately, they should be stored appropriately at refrigeration temperature.
- Immerse peeled or sliced light-colored fruits and vegetables (apples, pears, mushrooms, potatoes, etc.) in ascorbic acid solution to prevent discoloration prior to the thermal process.
- Fill hot foods into jars and adjust headspace as specified in recipes. Headspace is the unfilled space above the food in a container and below its lid (Figure 10.5). Directions for canning specify leaving ½ inch for fruit and tomatoes to be processed in boiling water, and from 1 to 1¼ inches in low-acid foods to be processed in a retort.[12] This space is needed for expansion of food as jars are processed, and for forming vacuums in cooled jars. The extent of expansion is determined by the air content in the food and by the processing temperature. Air expands greatly when heated to high temperatures; the higher the temperature, the greater the expansion. Foods expand less than air when heated.
- Tighten screw bands securely for glass jars, but not too tightly.
- Process and cool jars as soon as possible.
- Store the jars in a cool, dry, dark place, preferably between 50 and 70°F (10 and 21°C).

FIGURE 10.5
Illustration of Headspace in Glass jars

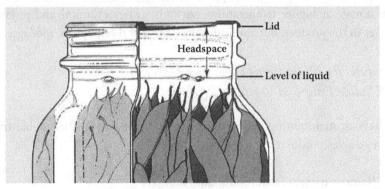

Source. Extracted from USDA, *Complete Guide to Home Canning*, Agriculture Information Bulletin No. 539, Revised 2009.[13]

Summary

Due to increased consumer awareness of the potential health benefits of fruits and vegetables, and their distinguished flavor and taste, the market demands on specialty fruit and vegetable products continuously grow. Some of the unique characteristics of specialty fruit and vegetable products are exotic fruit and vegetable varieties, organically or locally grown, processing technologies applied for retaining the bioactive compounds, color and flavor while ensuring shelf-life, preservative-free, etc. Different Forms of products may be processed, including frozen, dehydrated, and shelf-stable. As for other specialty foods, premium quality, unique taste and flavor, attractive package, and reasonable price are some of the keys for the success in this business.

References

1. Tanner, R. 2011. *The state of the specialty food industry 2011*. National Association for the Specialty Food Trade (NASFT).
2. Beale, B. 2011. *Growing specialty vegetables in Maryland: A growing field*. University of Maryland Extension Service publication. http://mredc.umd.edu/hotlistSpecialtyVegetables.htm (accessed October 1, 2011).
3. Fennema, O. 1977. Loss of vitamins in fresh and frozen foods. *Food Technology* 12:32–38.
4. Zhao, Y. 2007. Freezing process of berries. In *Berry fruit: Value-added product for health promotion*, ed. Y. Zhao, 291–312. Boca Raton, FL: CRC Press, Taylor & Francis Group.
5. Barbosa-Cánovas, G.V., Altunakar, B., and Mejía-Lorio, D.J. 2005. Freezing of fruits and vegetables. An agri-business alternative for rural and semi-rural areas. Rome: Food and Agriculture Organization of the United Nations. http://www.fao.org/docrep/008/y5979e/y5979e00.htm (accessed October 1, 2011).
6. Pham, Q.T., and Mawson, R.F. 1977. Moisture migration and ice recrystallization in frozen foods. In *Quality in frozen foods*, ed. M.C. Erickson and Y. Hung. New York: Chapman & Hall.
7. Dauthy, M.E. 1995. *Fruit and vegetable processing*. FAO Agricultural Services Bulletin 119. Rome: Food and Agriculture Organization of the United Nations. http://www.fao.org/docrep/V5030E/V5030E00.htm (accessed October 1, 2011).
8. Arthey, D.A., and Ashurst, P.R. 2001. *Fruit processing. Nutrition, products, and quality management*. Gaithersburg, MD: Aspen Publishers.
9. Fellows, P. 2004. *Small-scale fruit and vegetable processing and products: Production methods, equipment and quality assurance practices*. Vienna: United Nations Industrial Development Organization.
10. Figuerola, E.F. 2007. Dehydration of berries. In *Berry fruit: Value-added products for health promotion*, ed. Y. Zhao, 313–334. Boca Raton, FL: CRC Press, Taylor & Francis Group.
11. Zhao, Y., and Xie, J. 2004. Practical applications of vacuum impregnation in fruit and vegetable processing. *Trends Food Sci. Technol.* 15(9): 434–451.

12. Ramaswamy, H.S., and Meng, Y. 2007. Commercial canning of berries. In *Berry fruit: Value-added products for health promotion*, ed. Y. Zhao, 335–366. Boca Raton, FL: CRC Press, Taylor & Francis Group.
13. USDA. *Complete guide to home canning*. Agriculture Information Bulletin 539, USDA revised 2009. http://www.uga.edu/nchfp/how/general.html (accessed October 1, 2011).

11

Specialty Entrées, Meals, Convenience Foods, Soups, and Miscellaneous

Jingyun Duan

Oregon Department of Agriculture
Salem, Oregon

Ḱ

Contents

Introduction

Specialty meals cover a wide range of specifically prepared foods that are gourmet, premium, healthy, and ready to serve. They not only exemplify quality and innovation by manufacturing with better ingredients or artisan techniques, but also provide convenience and bring exotic flavor and a healthier lifestyle to the consumers.

According to the report "Today's Specialty Food Consumer 2010" by the National Association for the Specialty Food Trade (NASFT), there is a growing consumers' preference for eating at home. Along with the improving economic conditions, there was a surprising number of consumers who had reengaged with specialty foods in 2010.[1] Consumers eat at home more and are looking for ingredients to "bring restaurant experiences" to their own kitchens. Meanwhile, since young consumers frequently have limited cooking skills and older consumers limited time, specialty meals, such as gourmet frozen/refrigerated entrées, pizzas, and convenience foods, are turning in torrid sales.

This chapter discusses the characteristics of specialty entrée, pizzas and convenience foods, specialty meat, poultry, and seafood products, specialty soups, and specific processing technologies and controls for ensuring quality and food safety of these products. Examples of different types of specialty entrée, pizzas, meat, poultry, seafood, and soups are also given in this chapter.

Specialty Entrées, Pizzas, and Convenience Foods

Prepackaged, refrigerated, or frozen ready-to-eat complete meals, typically featuring a main dish with some sides, are gaining popularity for their convenience. Refrigerated ready meals continue to be the option of busy professionals who have little time for cooking but still seek the freshness offered by refrigerated meals.[2] Meanwhile, frozen ready meals, perceived as inexpensive and high in quality, account for the highest share (40%) in overall frozen food segment due to their wide product range, from frozen entrée to desserts and snacks.[3] It is reported that specialty frozen pizzas, which offer unique crusts, styles, and flavor combinations that go beyond traditional offerings and help achieve an image of restaurant quality, are driving value growth in the United States.[2]

Characteristics of Specialty Entrées, Pizzas, and Convenience Foods

The specialty entrées, pizzas, and convenience foods are generally made from higher-quality, all-natural, or organic ingredients, or with artisan techniques. The major characteristics of these specialty meals may include the following:

- Bring exotic flavors within ethnic cuisines. With Americans always ready to discover international dishes and flavors, exotic exploitation is evergreen. It is reported that 7 out of 10 shoppers aged 25 to 39 purchase ethnic foods at least once a month, with Chinese, Italian, and Mexican foods being the top 3 leading ethnic choices.[4]
- Address the health and diet concern. Consumers are increasingly interested in eating a healthy diet. To live a healthy lifestyle, many consumers are reading labels and avoiding foods that contain high calories, saturated fats, trans-fats, cholesterol, salt (sodium), and added sugars.[5] In addition to low calories and light food, consumers are also looking for food containing high potassium, fiber, vitamins A and C, calcium, and iron to stay fit and healthy and reduce the risk of many chronic diseases.
- Provide natural food that contains no artificial ingredients and chemical preservatives. Natural food, defined as food not containing any artificial flavoring, coloring ingredient, or chemical preservative, or any other artificial or synthetic ingredients, and in the case of meat and poultry, is minimally processed, is a growing trend in the marketplace.[6] Consumers of natural food acknowledge that they eat these products for health and wellness reasons, but also because they taste better.
- Fulfill the gluten-free needs. In the National Restaurant Association's annual survey for the top 20 trends for 2011, gluten-free food was number 8 on the list, up one spot from 2010.[7] With an estimated 1 in every 105 Americans affected by celiac disease,[8] a medical condition in which the absorptive surface of the small intestine is damaged by gluten, there has been a strong push to offer more gluten-free food products. Gluten is a protein found in wheat, rye, triticale, and barley. In addition to improving the overall quality of life for people with celiac disease, a gluten-free diet may also improve other chronic health issues.[6] The gluten-free foods are expanding to a much larger audience who wants to eat healthier and better-for-digestion foods. The gluten-free market has grown at an average annual rate of 28% since 2004.[9]

According to Tanner,[1] specialty brands of entrées, pizzas, and convenience foods may include Tandoor Chef, Ethnic Gourmet, Bertolli, Vicolo, José Olé, Lean Cuisine, Healthy Choice, Kashi, Bell & Evans, Dr. Praeger's, AllergyFree Foods, Feel Good Foods, and Conte's Pasta. Table 11.1 gives examples of some specialty entrées, pizzas, and convenience meals on the U.S. market.

Quality and Food Safety Considerations of Refrigerated and Frozen Meals

Refrigeration and freezing are the major forms of food preservation used for specialty entrées, pizzas, and convenience meals today. Refrigeration retards many of the microbial, physical, chemical, and biochemical reactions that lead to food spoilage and deterioration, thus providing effective short-term food preservation. Freezing preserves food for extended periods by slowing the movement of molecules and causing microorganisms to enter a dormant stage.[10] However, keeping food refrigerated or frozen does not kill bacteria, and temperature abuse can easily occur, especially in refrigerated food. Having safe and high-quality refrigerated and frozen foods requires a low initial bacteria count, minimal contamination (including cross-contamination), and rapid chilling/freezing during processing, and low temperatures during storage, handling, distribution, retail display, and consumer storage.[11]

Effective control of the "cold chain," a term used to describe the series of interdependent operations in the production, distribution, storage, and retailing of chilled and frozen foods, is vital to ensure food safety and quality of refrigerated and frozen foods. The following are the key recommendations for both refrigerated and frozen foods:[11]

- Maintain good conditions of hygiene at all stages of food processing, distribution, and storage.
- Chill or freeze products rapidly and adequately after preparation.
- Maintain chill (<5°C) or frozen (<−18°C) temperatures, wherever possible, during storage, distribution, and retail display.
- Ensure that chilled or frozen products are transferred in a continuous operation with no stopping or delays between temperature-controlled areas, e.g., delivery trucks to holding stores, storage areas to retail display units.
- Separate cooked and uncooked chilled or frozen products in storage and retail display areas to prevent cross-contamination.
- Conduct frequent and systematic temperature checks on product temperatures using appropriate and calibrated instrumentation.

TABLE 11.1
Examples of Some Specialty Entrées, Pizzas, and Convenience Foods

Exotic Products

Tandoor Chef© Lamb Vindaloo with turmeric-infused basmati rice: Rich in Indian flavor, made fresh from all-natural ingredients, and instantly frozen to lock in natural nutrients, flavors, and color.

Bertolli© Chianti Braised Beef and Rigatoni: Made from fresh, simple ingredients and extra virgin olive oil and can provide authentic Italian fine dining experience in minutes with one skillet or by oven/microwave.

Vicolo© Corn Meal Crust Pizza: Made from non-GMP organic corn meal and topped with freshly prepared seasonal ingredients to create a distinctively Italian taste and texture.

José Olé© Chicken Taquitos Corn Tortillas: Made from finest ingredients, like tender cuts of meats, real cheese, oven-baked tortillas, and authentic seasonings and spices to offer a genuine Mexican flavor.

Gluten-Free Products

Bell and Evans© Gluten Free Breaded Chicken Breast Nuggets: Whole breast meat pieces are lightly breaded with rice flour for a great taste.

Dr. Praeger© Gluten Free California Veggie Burgers: Provide great taste without using any wheat, oats, matzoh meal, or bran.

Feel Good Foods© Handmade Gluten Free Asian Style Dumplings: All-natural ingredients with no MSG, artificial flavors, preservatives, or additives, accompanied by a gluten-free dipping sauce.

Conte© Margherita Gluten-Free Pizza: Contains award-winning tomato sauce and the freshest cheese and dough.

All-Natural Products

Healthy Choice© Asian Potstickers: A delectable vegetarian dish served on a healthy bed of whole-grain rice and covered with a sweet Asian-style sauce, made with premium, enticing, all-natural ingredients.

Kashi© Thin Crust Mediterranean Pizza: Made from all-natural ingredients, the crust is made with seven whole grains and sesame with flax seed and topped with fresh and wholesome ingredients—roasted vegetables, unique mushrooms, and tasty cheeses.

Healthy Products

Lean Cuisine© Shanghai-Style Shrimp: Contains low calories (250 calorie) and low fat and cholesterol.

Healthy Choice© Chicken Pesto Alfredo: A balanced meal containing chicken tenderloins in basil-pesto Alfredo sauce with whole-grain pasta, vegetables, cherry, and blueberry multigrain crisp, high in fiber and antioxidants, and low in fat.

Temperatures can be measured directly by contacting with the food or indirectly by measuring the environment or between packages.

- Do not overload chilled or frozen storage and retail areas with product; allow sufficient air circulation in storage and retail areas.
- Train and educate all personnel to follow the correct handling and storage of chilled and frozen foods.

Specialty Meat and Poultry Products

Specialty meat and poultry products may include fresh sausage, cooked hams, smoked poultry, and pickled and cured meats. The possibilities of specialty meat products are endless. They may be served as a handy dish if consumers need an entrée that is quick and easy, but nutritious, and also as a lavish dish if elegantly prepared.[12]

Exotic and Ethnic Meat and Poultry Products

Beef, pork, chicken, and turkey are the most popular types of meat and poultry consumed throughout the world today. Some types of meat, such as lamb, venison, buffalo, elk, and duck, when made into meat products, are considered to be more specialized and exotic because of the small amount with regard to consumption. For example, merguez, a red, spicy lamb-based fresh sausage, is made with lamb, beef, or a mixture stuffed into a lamb intestine casing and heavily spiced with chili pepper or harissa to give a characteristic piquancy and red color. It is a North African cuisine, and also popular in France, Israel, the Middle East, and Northern Europe.

Development of more ethnic/international flavors/varieties is another growth opportunity for specialty meat products. Southeast and East Asia, and in particular China, have a rich tradition and long history in further processing of meat. Europe is also famous for a great variety of processed meat products with its own flavor and character.

Most traditional Asian meat products are fermented for extending shelf life and achieving desired flavor and taste. The typical characteristic of most products is the utilization of sugar as an ingredient for

- Achieving slight to moderate sweet taste commonly desired in Asia
- Lowering water activity in the presence of sugar and improving bacterial stability

- Enhancing fermentation processes, as sugars serve as "food" for fermentation bacteria (only in the case of fermented products)

The most popular product available throughout East and Southeast Asia is the Chinese sausage, a product neither fermented nor ripened. The principal ingredients of fresh raw pork and pork fat are mixed with nonmeat ingredients, including curing salt, sugar, pepper, garlic, and optionally, some Chinese seasonings, including cinnamon, ginger, soy sauce, and Chinese rice wine, and filled into pig casings or more recently into collagen casings. The products are subjected to a two-stage drying: a first stage of 2 days to temperature ~60°C, usually produced by charcoal (alternatively wood and electricity), and a second stage of 2–3 days at ~50°C. Dry heat is essential in Chinese sausage manufacture, and the flavor results basically from the ingredients used. There is also the option of replacing some of the meat and fat by approximately 20% pork liver.[13]

Shelf-stable fermented meat products are typically processed in Europe. Examples are as follows:

- Raw-cured ham: Prosciutto di Parma (Italy), Jamon serrano (Spain), Bayonne ham (French), and Black Forest ham (Germany).
- Dry-fermented sausage: Hungarian salami, Italian salami, Spanish chorizo, and French saucisson.

Raw-fermented meat products (sausages) receive their unique properties, such as tangy flavor, chewy texture, and intense red curing color, through several months of fermentation processes.[14] In addition to fermentation, ripening phases combined with moisture reduction are necessary to build up the typical flavor and texture of the final product. Air-dried fermented meat products are made throughout Southern Europe, while the ones from Central and Northern Europe are in many cases cold-smoked. The time-consuming manufacture of raw fermented hams and sausages is done mainly in the cold winter season, as relatively low temperatures are required for fermentation, drying, and ripening.[14]

Convenient Meat and Poultry Products

Convenience is also a critical component of success in marketing meat products. Resealable packaging, such as press-to-close or slide zipper, as well as reusable plastic containers with closable lids, is one of the latest innovations, particularly in the lunch meat segment. The deli-style lunch meat available in thinly shaved and thick-sliced varieties, combined with the innovative packaging, would lure consumers away from the deli counter by offering them products "without the wait."

Resealable packaging yields a consumer-friendly, easy-open, and reclosable package, which has many advantages, including consumers can easily open the package without the use of a knife or other sharp utensil, food stays fresh over a long period even if the product is used multiple times from the opening, and possible leakage and contamination are avoided. Specialty meat and poultry producers should continually explore new technologies and packaging solutions to meet consumers' increasing demand for convenience and sustainability.

Healthy and All-Natural Meat and Poultry Products

The healthier eating trend has also had an impact on the meat product market. Product launches included reduced-fat, reduced-sodium, or better-for-you variants, such as turkey bacon or lower-fat sausage.[15] Many meat processors claim that their products are minimally processed with no artificial ingredients. Balanced meat products, in which fruits and vegetables are added for novel flavor, are also attracting consumers and increasing usage occasions.

"Naturally cured" and "nitrite-free" meat products have appeared on the market to promote a health image to consumers. Many meat processors also claim that their animals are raised naturally and humanely, and all their livestock is never given antibiotics or any added hormones, which is generally believed to be better for the animals, the environment, and ultimately consumers.

The USDA *Food Standards and Labeling Policy Book* defines *natural* of processed meat and poultry products as "the product does not contain any artificial flavor or flavoring, coloring ingredient, or chemical preservative (as defined in 21 CFR 101.22), or any other artificial or synthetic ingredient; and the product and its ingredients are not more than minimally processed."[16] Conventionally cured meat products are characterized by addition of nitrite, which contributes to the development of the specific cured flavor and color. Nitrite acts as an antimicrobial agent to retard the germination of spores and toxin formation by *Clostridium botulinum*, and prevents lipid oxidation by its strong antioxidant property.[17] However, nitrite is also responsible for the formation of carcinogenic N-nitrosamines in some cured products under certain processing conditions, and therefore "naturally cured" and "nitrite-free" meat products have enjoyed a widespread market acceptance.

According to a survey on 56 commercial "uncured" meat products, including bacon, ham, frankfurters, bologna, braunschweiger, salami, Polish sausage, Andouille sausage, and snack sticks, most of these products demonstrated typical cured meat color and appearance by adding sea salt, evaporated cane juice, raw sugar or turbinado sugar, lactic acid starter culture, natural spices or natural flavorings, and celery juice or celery juice concentrate.[18] Sea salt is the most common ingredient observed in the product labels of natural processed meats. Although sea salt has been suggested as a likely source of nitrite, studies suggested that nitrite content of sea salt is relatively low.[19] The second most common ingredient in natural processed meat products is turbinado sugar, which is a raw sugar obtained from evaporation of sugar cane juice followed by centrifugation to remove surface molasses. Natural flavorings or spices, and vegetable juice, such as celery juice, are frequently used as ingredients in naturally cured meat products. Vegetables are well known as a source of nitrate.[20] Nitrate can be reduced to nitrite and enter into curing reactions by microorganisms found in the natural flora of meat or by intentional addition of microorganisms with nitrate-reducing properties. Cultures of coagulase-negative cocci, such as *Kocuria varians*, *Staphylococcus xylosus*, and *Staphylococcus carnosus*, can achieve nitrate reduction at 15–20°C, but are much more effective at temperatures over 30°C.[18] It has been reported that the typical recommended holding temperature for commercial nitrate-reducing cultures is 38–42°C to minimize the time necessary for adequate nitrite formation, and time is a critical parameter in the development of typical cured meat properties from natural sources of nitrate. The relatively small amounts of nitrite formed from natural nitrate sources provide a sufficient antioxidant role in natural processed meats as typically observed in nitrite-cured meats.

Examples of some specialty meat and poultry products are illustrated in Table 11.2.

Specialty Seafood

Specialty food consumers are now turning their attention to seafood, which is considered to be healthier than red meat and other sources of protein because of its low fat content. Seafood contains high-quality proteins and a number of essential nutrients, and many species are full of omega-3 fatty acids that have been found to help prevent heart disease, cancer, autoimmune disease, brain aging, and Alzheimer's disease.[21]

TABLE 11.2
Examples of Some Specialty Meat and Poultry Products

Exotic and Ethnic Products

Fabrique Delices© Merguez Sausage: A spicy Mediterranean all-natural lamb and beef sausage perfect on the grill.	Asian Taste© Chinese Style Sausage: A dried, hard sausage made commonly from pork and a high content of fat, and normally smoked, sweetened, and seasoned.	Volpi Foods© Sliced Coppa: Originated from Italy; fully air-dried to ensure a delicate flavor.	Volpi Foods© Traditional Italian Prosciutto Ham: The product of a long, unhurried drying process lasting a minimum of 210 days.

Convenient Products

Danola© Supreme Smoked Maple Flavored Honey Ham: Packaged in a resealable Zip-Pak.	Oscar Mayer© Mini Beef Hot Dogs: Packaged in a convenient, stand-up refrigerator pouch with an easy-to-use, resealable zipper.	Sara Lee© Oven Roasted Turkey Breast: Presliced and packed in a press-to-close resealable plastic package.	Oscar Mayer© Deli Fresh Shaved Ham: Sliced, packed, and sealed in a reclosable and reusable plastic container.

Healthy and All-Natural Products

Jennie-O© Extra Lean Turkey Bacon: 95% fat-free; made from lean turkey, a delicious alternative to pork bacon.	Aidells© Spicy Mango with Jalapeño Smoked Chicken Sausage: Made from all-natural ingredients, including tender poultry with sweet mango, spicy jalapeño, cilantro, and paprika.	Al Fresco© Tomato and Basil Chicken Meatball: Absolutely no artificial ingredients; 60% fat-free, gluten-free, and pork-free.	Niman Ranch© Center Cut Bacon: Hand-rubbed with spices, center-cut and slow-smoked over sweet applewood chips. The animals are humanely raised on sustainable U.S. family farms and ranches, and never given antibiotics or hormones.

Trends in Seafood Products

Wild, Natural, and Sustainable Seafood Products   With a growing interest in wild, natural, and sustainable seafood choices, consumers are requesting more information and expecting the seafood industry to meet these new, higher expectations.[22] Many seafood manufacturers have answered the call and helped support sustainable fisheries by providing high-quality, healthy, and all-natural seafood from environmentally responsible and credible sources. For example, many fine quality seafood items are low in fat, sodium, cholesterol, and contain no sugars or unnecessary additives or preservatives. There are also various seafood products that contain no preservatives, dyes, growth hormones, antibiotics, pesticides, moisture retainers, carbon monoxide, tasteless smoke, etc., and are sourced exclusively from environmentally sustainable fisheries.

Value-Added Seafood Products   Fueled by changing lifestyle, consumers have a growing appetite for value-added seafood products. For consumers, value-added seafood implies convenience, meaning the purchased products have had something done to them to take the effort away from the end user.[23] For seafood manufacturers, particularly small seafood processing firms that suffer from seasonal inactivity when harvesting is officially closed in the winter, successfully innovated value-added processing ensures year-round activity and brings maximized profit.[24]

Smoked fish or shellfish is considered a luxury and often sold in gourmet shops. The smoking process was originally used as a preservative, but in more recent times the smoking of fish is generally done for the unique taste and flavor imparted by the smoking process. The variables of temperature, humidity, and characteristics of the wood and raw fish all must be taken into account and kept in balance. Fish can be cold smoked at temperatures below 26°C for several days to take on a smoked flavor, but remain relatively moist. Cold smoking does not cook food, but hot smoking with temperatures up to 121°C for several hours fully cooks the fish and imparts fish the smoke flavor. Hot-smoked fish is typically safe to eat without further cooking. Any fish and most kinds of wood will work for smoking fish. However, fattier fish, such as salmon and trout, will absorb more smoke flavor and therefore are perfect for smoking. Woods like alder or fruit woods are traditionally used in fish smoking. Mesquite, oak, or any favorite smoke maker will also give fish a distinguished flavor if well controlled.[25]

Salt-cured fish is another value-added seafood product. Salting, with either dry salt or brine, inhibits the growth of microorganisms by drawing water out of microbial cells through osmosis, firms the protein for good texture, and deactivates some enzymes.[26] At the same time, the salt intensifies the natural flavor of the product. Often the brine would contain spices other than salt to add extra flavor to the fish. Once cured, the fish can serve as a deeply flavorful base for a dish. For example, salt-cured anchovies are available as flat fillets or rolled fillets with capers, and their unique flavor is perfect for salads, casseroles, appetizers, and pizza.

Luxury Seafood Luxury seafood, such as crabmeat, jumbo shrimp, lobster tails, caviar, and escargot, is the symbol of the specialty food business. With recent developments in aquaculture, combined with American ingenuity, these luxury foods are increasingly accessible to all consumers.

Caviar is the processed, salted, and unfertilized roe of the sturgeon fish. However, in the United States nonsturgeon roe, such as salmon, steelhead, trout, lumpfish, and whitefish roe, has also been accepted as caviar as long as the source fish precedes the word *caviar*. For instance, the roe of farmed rainbow trout is called plainly trout caviar.[27] The malossol (means lightly salted) process is a preferred processing method by experts in making caviar. This type of caviar usually has a maximum of 5% salt and is considered a high-quality type. Semipreserved or salted caviar has a salt content of 8%. It has a longer shelf life due to the high salt content, but its taste is compromised. Pressed caviar, made from overly ripe roe, has a high salt content and is pressed to a consistency like jam. It is one of the favorite types of caviars due to its strong and concentrated flavors. Pasteurized caviar is made from fresh caviar. It is treated in high temperature and vacuumed packed to be put in glass jars packed for longer preservation. Although taste and texture are usually compromised, it is the improved process of the traditional method, where it can lengthen the shelf life of the caviar.[28]

Examples of some specialty seafood products are shown in Table 11.3.

Seafood HACCP

Seafood safety is a growing concern. The potential hazards related with seafood and seafood processing are listed in Table 11.4. Seafood producers shall develop and implement evidence-based seafood safety controls related specifically to their kinds of seafood and seafood operations.

To respond to the increased concern over seafood safety, the FDA has issued regulations that mandate the application of Hazard Analysis and Critical Control Point (HACCP) principles to the processing of seafood (21 CFR 123). The FDA has also published "Fish and Fishery Products Hazards and Controls Guidance" to assist processors of fish and fishery products in the development of their HACCP plans.[29] This guidance document provides a series of 18 steps (see Table 11.5) that will yield a completed HACCP plan. Fish and fishery processors could follow these steps to identify the physical, chemical, and biological hazards that are associated with their products, and formulate control strategies that cover all aspects of food production, from raw materials, processing, distribution, and point of sale, to consumption and beyond, to ensure the safety of their products.

TABLE 11.3
Examples of Some Specialty Seafood Products

Wild, Natural, and Sustainable Products

EcoFish© Battered Wild Alaskan Salmon Fillets: Premium-quality wild Alaskan salmon fillets with special crispy all-natural gluten-free batter and sourced from environmentally sustainable fisheries.

Crown Prince© Fancy White Lump Crab Meat: Does not contain any MSG or sulfites, and no bleaching agents or preservatives are used in the rinsing, washing, or canning process.

Crown Prince© Skinless and Boneless Pacific Pink Salmon: Contains 235 mg of omega-3 fatty acids per serving; is created from wild-caught fish from sustainably harvested salmon fisheries off the Pacific Northwest Coast.

Value-Added Products

St. James© Scotch Reverse Cold Smoked Salmon: Gently cured and smoked over Scottish whisky oak chips, with a distinct and bold smoke flavor.

Ekone© Smoked Oysters: Made with fresh ingredients and no preservatives, and have five great flavors for making a great gift.

Martel© Flat Fillets of Anchovies: Preserved in salt and olive oil; flavorful, meaty, and boneless.

Crown Prince© Cured Anchovies: Cured with kosher salt and packed in pure olive oil.

Luxury Products

Maryland© Jumbo Lump Crab Meat: Contains only blue crab meat and has a distinctly different taste.

California Caviar Company© Pressed Caviar: Combines the nutty flavor of California white sturgeon, the assertive full sea flavor from paddlefish, and the creamy, rich taste of hackleback roe.

Marky© Salmon Keta Caviar: The select roe of Alaskan chum salmon has a strong, sweet, and honey-like flavor and a golden-orange color.

Specialty Soups

Usually as a companion to sandwiches, soup is also becoming a year-round one-dish meal, showing a continued growth in the market, with innovation being a key to boosting consumer interest.[30]

Characteristics of Specialty Soups

The majority of consumers consider soup a comfort food, but few people have the time to make soup stocks from scratch. Therefore, convenience and ease of preparation continue to dominate the product lines and packaging for soup products.

TABLE 11.4
Potential Seafood Safety Hazards

Potential Vertebrate Species-Related Hazards

- Parasites
  - Roundworms: e.g., *Anisakis* spp., *Pseudoterranova* spp., *Eustrongylides* spp., and *Gnathostoma* spp.
  - Tapeworms: e.g., *Diphyllobothrium* spp.
  - Flukes: e.g., *Chlonorchis sinensis, Opisthorchis* spp., *Heterophyes* spp., *Metagonimus* spp., *Nanophyetes salmincola,* and *Paragonimus* spp.
- Natural toxins
  - Paralytic shellfish poisoning (PSP), neurotoxic shellfish poisoning (NSP), diarrhetic shellfish poisoning (DSP), amnesic shellfish poisoning (ASP), ciguatera fish poisoning (CFP), and azaspiracid shellfish poisoning (AZP)
- Scombrotoxin (histamine)
- Environmental chemicals: e.g., industrial chemicals (including heavy metals and pesticides)
- Aquaculture drugs

Potential Invertebrate Species-Related Hazards

- Pathogens
  - Pathogens of human or animal origin: e.g., *Vibrio cholerae* O1 and O139, *Salmonella* spp., *Shigella* spp., *Campylobacter jejuni, Yersinia enterocolitica,* hepatitis A virus, and norovirus
  - Naturally occurring pathogens: e.g., *Vibrio vulnificus, Vibrio parahaemolyticus,* and *V. cholerae* non-O1 and non-O139
- Parasites
- Natural toxins
- Environmental chemicals
- Aquaculture drugs

Potential Process-Related Hazards

- Pathogenic bacteria growth-temperature abuse
- *Clostridium botulinum* toxin
- *Staphylococcus aureus* toxin
- Pathogenic bacteria survival through cooking or pasteurization
- Pathogenic bacteria survival through processes designed to retain raw product characteristics
- Pathogenic bacteria contamination after pasteurization and specialized cooking processes
- Allergens/additives
- Metal inclusion
- Glass inclusion

Source. Adapted from U.S. Food and Drug Administration, Fish and Fishery Products Hazards and Controls Guidance, 4th ed., 2011, http://www.fda.gov/downloads/Food/GuidanceCompliance Regulatory Information/GuidanceDocuments/Seafood/UCM251970.pdf.

TABLE 11.5
Eighteen Steps to Develop a Complete HACCP Plan

Preliminary Steps

- Provide general information.
- Describe the food.
- Describe the method of distribution and storage.
- Identify the intended use and consumer.
- Develop a flow diagram.

Hazard Analysis Worksheet

- Set up the hazard analysis worksheet.
- Identify potential species-related hazards.
- Identify potential process-related hazards.
- Understand the potential hazard.
- Determine whether the potential hazard is significant.
- Identify critical control points.

HACCP Plan Form

- Set up the HACCP plan form.
- Set critical limits.
- Establish monitoring procedures (what, how, frequency, who).
- Establish corrective action procedures.
- Establish a recordkeeping system
- Establish verification procedure.
- Complete the HACCP plan form.

Source. Adapted from U.S. Food and Drug Administration, Fish and Fishery Products Hazards and Controls Guidance, 4th ed., 2011, http://www.fda.gov/downloads/Food/GuidanceCompliance Regulatory Information/GuidanceDocuments/Seafood/UCM251970.pdf.

Microwavable containers and one-cup portable portions are giving a realistic option for people eating on the go. These ready-to-serve products make soup faster, easier, and simpler to handle, widen the category's appeal for consumers, and provide consumers the option for enjoying shippable soup anywhere and anytime.

Instant soups also provide conveniences for consumers as an alternative to vending machine lunches or fast food. Packaged instant soups are available in several varieties, such as a packet of dry soup stock, and miso paste soup packed in foil-lined pouches with preserved vegetable condiments. These soups are ready in several minutes by adding hot water or by adding water and heating for a short time.

Consumers' desire for more healthful food has an impact on the soup market as well. Canned soups used to get a bad reputation for high sodium content. New, healthier products have addressed the issue with reduced-sodium options.[31] In addition to low-sodium products, all-natural ingredients and low-carb soups also have been taking more space on store shelves. The health claims for these specialty soups include the following:

- Low sodium
- Low fat/cholesterol or fat/cholesterol-free
- High fiber
- Low calories
- All-natural ingredients

A successful healthy soup should exploit consumer interests to open the door to higher-value soups.

Although soups are tied to tradition and provide a level of comfort that consumers identify with, soups with an international bent are also garnering interest.[32] Americans are actively seeking out new flavors from many different cultures. Asian food is one of the fastest-growing segments of ethnic soups. Consumers enjoy Asian flavors with selections such as Korean kimchi soup, Japanese miso soup, Thai Tom Yum Soup, and Chinese hot and sour soup. Latin soups, such as roasted pumpkin soup and tortilla soup, also have a lot of potential. Italian soups, such as Sicilian pasta fagioli soup, are another area that is growing in popularity.

Examples of some specialty soup products are shown in Table 11.6.

Package Design of Soup Cups

Prepackaged instant soups are a frequent cause of burn injury, especially in children.[33] It is reported that package design of soup cups may increase the risk for burn injury by affecting container stability.[34] The ease of tipping over is correlated with the cup base area, top area, and ratio of height/base area, and the most significant contributor to the ease of tipping over is height. Soup cups that are tall with a narrow base are predisposed to being knocked over and spilled. Designing of instant soup packaging with a wider base and shorter height, along with the requirement for warnings about the risks of burns, would reduce the frequency of soup burns and ensure the safety of consumers.

Summary

Innovation is always the key to boosting consumer interest in the specialty food market. Beyond overt claims of convenience/ease of preparation, the

TABLE 11.6 Examples of Some Specialty Soup Products			
Convenient Products			
Campbell© Tomato Soup: Offers great-tasting soup flavors in convenient microwavable bowls.	Campbell© Soup at Hand New England Clam Chowder: The heat-and-go cups provide consumers the option for enjoying shippable soup anywhere and anytime.	Nile Spice© Country Mushroom Soup: All-natural soup that is ready in 5 minutes after adding hot water.	Kikkoman© Tofu-Spinach Miso Soup: Contains powdered red and white miso, and dehydrated tofu, seaweed, and spinach.
Healthy Products			
Campbell© Healthy Request Old Fashioned Vegetable Beef Soup: Highlights the natural sea salt in its low-sodium, 98% fat-free soup.	Progresso© Reduced Sodium Vegetable Soup: Light and high-fiber soup with 80 calories or less.	Pacific Natural Foods© Organic Free Range Chicken Broth: Made with low-sodium, grain-fed chicken, contains no hormones/ antibiotics, and uses organic seasonings.	Wolfgang Puck© All-natural Classic Tomato Basil Soup: Restaurant-quality premium soups made with finest all-natural ingredients.
Ethnic Products			
Annie Chun© Chinese Hot and Sour Soup Bowl: Captures traditional Chinese flavors, 100% natural, and has no MSG or preservatives.	Nueva Cocina© Roasted Pumpkin Soup with Ginger: Combines the sweet taste of pumpkin with subtle notes of ginger and cinnamon, made from all-natural ingredients with no MSG and no preservatives.	That's A Nice© "Zuppa Tomato" Mamma's Sicilian Pasta Fagioli Soup: An original and unique blend of white cannellini beans and pasta in a rich tomato base, creating an authentic home-style Sicilian hearty ready-to-serve soup.	

market has mostly been driven by lifestyle choices that impact attitudes toward health and wellness. The "no additives/preservatives" claim is the most common trend. Additionally, better-for-you claims, including low/no/ reduced-fat/trans-fat/calorie jumped in the past year.[15] For specialty meal manufacturers, promptly realizing these trends and making smart responses in the production lines lead to a successful market.

For certain specialty meals, some uncommon and poorly characterized ingredients and processing procedures may be used in the production, which may lead to potential allergen or microbial safety problems. The Food Allergen Labeling and Consumer Protection Act of 2006 (FALCPA) requires packaged foods to clearly label all ingredients that are class I allergens, including peanuts, tree nuts, milk, eggs, soy, wheat, seafood, and crustaceans (the "big 8" food allergens), and against which may lead to class I food recalls. If considering export to other countries, it is important to note that other countries may have additional labeling requirements for allergen disclosure. For example, in Canada sesame seeds and more recently mustard seeds are now considered food allergens. In addition, nutritional labeling has become an important part of a product's marketing mix since consumers today routinely review the nutritional contents of a product as part of their decision-making process.[35]

Appropriate quality control operations shall be employed to ensure that the ingredients to be incorporated into finished food are premium and consistent. All reasonable precautions shall be taken to ensure that production procedures do not contribute contamination from any source. The implementation of effective food safety programs is essential for ensuring safe production of specialty meals.

References

1. Tanner, R. 2010. Today's specialty food consumer 2010. *Specialty Food Magazine*, C2.
2. Redruello, F. 2010. Ready meals flourish in the midst of economic uncertainty. http://blog.euromonitor.com/2010/11/ready-meals-flourish-in-the-midst-of-economic-uncertainty.html.
3. Business Wire. 2011. Research and markets: Global frozen food market analysis by products type and by geography—Trends and forecasts (2010–2015).
4. Shoukas, D. 2011a. The exploding American palate: Ethnic flavors spice up the mix. http://www.specialtyfood.com/news-trends/featured-articles/market-trends/the-exploding-american-palate-ethnic-flavors-spice-up-the-mix/.
5. Frazier, K. 2011. Trend for eating healthy food. http://diet.lovetoknow.com/wiki/Trend_for_Eating_Healthy_Food.
6. USDA Foreign Agricultural Service. 2010. Canada trends—Natural/health foods. http://www.b-for.com/Forms/CanadaTrendsNaturalHealthFoods.pdf.
7. National Restaurant Association. 2010. Chef's survey: What's hot in 2011. http://www.restaurant.org/pdfs/research/whats_hot_2011.pdf.
8. Rewers, M.J. 2005. Epidemiology of celiac disease: What are the prevalence, incidence, and progression of celiac disease? *Gastroenterology* 128(4 Suppl 1): S47–51.
9. Shoukas, D. 2011b. Gotta have gluten-free. http://www.specialtyfood.com/news-trends/featured-articles/retail-operations/gotta-have-gluten-free/.

10. U.S. Department of Agriculture. 2010. Freezing and food safety. http://www.fsis. usda.gov/factsheets/focus_on_freezing/index.asp.
11. UC–Davis. 2011. Managing the cold chain for quality and safety. http:// seafood.ucdavis.edu/pubs/coldchain.doc.
12. Barr, D. 2010. Specialty meats. http://liberianobserver.com/node/9262.
13. Heinz, G., and Hautzinger, P. 2007a. Traditional/ethnic meat products. http:// www.fao.org/docrep/010/ai407e/AI407E17.htm.
14. Heinz, G., and Hautzinger, P. 2007b. Raw-fermented sausages. http://www.fao. org/docrep/010/ai407e/AI407E11.htm.
15. Browne, D. 2011. Convenience meals and processed meat. Prepared food. March 2011,pp.51–58.http://www.businesswire.com/news/home/20110214006223/ en/Research-Markets-Global-Frozen-Food-Market-Analysis.
16. U.S. Department of Agriculture. 2005. *Food standards and labeling policy book.* www.fsis.usda.gov/OPPDE/larc/Policies/Labeling_Policy_Book_082005.pdf.
17. Shahidi, F., and R.B. Pegg. 1992. Nitrite-free meat curing systems: Update and review. *Food Chemistry* 43(3): 185–91.
18. Sebranek, J.G., and J.N. Bacus. 2007. Cured meat products without direct addition of nitrate or nitrite: What are the issues? *Meat Science* 77(1): 136–47.
19. Herrador, M.A., A. Sayago, D. Rosales, and A.G. Asuero. 2005. Analysis of a sea salt from the Mediterranean Sea. *Alimentaria* 360: 85–90.
20. Sindelar, J.J., J.C. Cordray, J.G. Sebranek, J.A. Love, and D.U. Ahn. 2007. Effects of varying levels of vegetable juice powder and incubation time on color, residual nitrate and nitrite, pigment, pH and trained sensory attributes of ready-to-eat uncured ham. *Journal of Food Science* 72(6): S388–95.
21. Das, U.N. 2008. Can essential fatty acids reduce the burden of disease(s)? *Lipids in Health and Disease* 7: 9.
22. Alaska Seafood Marketing Institute (ASMI). 2009. Menu Alaska. Consumer research results: A clear preference for wild, natural and sustainable Alaska seafood. http://www.alaskaseafood.org/foodservice/consumer-research/pdf/ Menu%20Alaska%20-%20Consumer%20Research.pdf.
23. Evans, J. 2005. Value added seafood trends. http://www.intrafish.no/global/ industryreports/article11316.ece.
24. Amankwah, F., K. Gunjal, P. Goldsmith, and G. Smith. 2003. Financial feasibility of producing value-added seafood products from shrimp water in Quebec. https://netfiles.uiuc.edu/pgoldsmi/www/articles_journals/2003-Aquatic.pdf.
25. Riches, D. 2011. Fish: Smoking. Your complete guide to smoking fish. http://bbq.about.com/cs/fish/a/aa030400a.htm.
26. Hilderbrand, K.S. 1993. Fish pickling for home use. http://osuext. intermountaintech.org/download/Fish%20Pickling%20for%20Home%20 use.pdf.
27. Maslow, J. 2011. Caviar: A luxury food in flux. http://www.specialtyfood.com/ news-trends/featured-articles/retail-operations/caviar-a-luxury-food-in-flux/.
28. Anonymous, 2011. Caviar processing methods. http://www.caviar-guide.com/ Caviar-Processing.htm.
29. U.S. Food and Drug Administration. 2011. Fish and fishery products hazards and controls guidance. 4th ed. http://www.fda.gov/downloads/Food/ GuidanceComplianceRegulatoryInformation/GuidanceDocuments/Seafood/ UCM251970.pdf.
30. Meszaros, E. 2011. Soup's up! http://www.specialtyfood.com/news-trends/ featured-articles/market-trends/soups-up/.

31. White-Sax, B. 2007. Convenient packaging, healthier profiles drive soup sales. http://www.drugstorenews.com/article/convenient-packaging-healthier-profiles-drive-soup-sales.
32. Denis, N.P. 2011. The new look of soups: Hearty, healthy, ethnic. http://www.specialtyfood.com/news-trends/featured-articles/retail-operations/the-new-look-of-soups-hearty-healthy-ethnic/.
33. Greenhalgh, D.G., P. Bridges, E. Coombs, D. Chapyak, W. Doyle, M.S. O'Mara, and T.L. Palmieri. 2006. Instant cup of soup: Design flaws increase risk of burns. *Journal of Burn Care and Research* 27(4): 476–81.
34. Greenhalgh, D.G. 2008. Pediatric soup scald burn injury: Etiology and prevention. *Journal of Burn Care and Research* 29(1): 114–18.
35. Gourmet Retailer. 2004. Major specialty food trends. http://www.allbusiness.com/retail-trade/food-stores/4217548-1.html.

12

Additional Information and Resources for Entrepreneurs

Yanyun Zhao

Oregon State University
Corvallis, Oregon

Ḱ

Contents

Introduction

There is no universal and standard definition on specialty foods, or different regulations from mainstream foods. The term *specialty foods* is rather used as a marketing term to promote the products due to their unique characteristics. Specialty foods are processed and regulated in the same fashion as other types of food products based on their acidity, water activity, etc. As described in the previous chapters of this book, specialty food products usually include artisanal, natural, and local products that are often made by small manufacturers, artisans, and entrepreneurs from the United States and abroad. Specialty foods may have one or more distinct qualities, such as natural or locally grown ingredients, unique processing procedures, manufactured on a small scale, exotic flavor and aroma, specific branding, distinctive packaging, ethnic foods, or distributed in specialty food retail stores or sections within supermarkets and grocery stores. Specialty food consumers have a general belief that specialty foods have a better taste and higher quality than mainstream and conventional ones. The retail sale of specialty food products continuously inclines annually, and had a total sale of $70.32 billion in 2010, accounting for 13.1% of all retail food sales.

Despite the continuous increase in the production of specialty food products, very few documents and literature are available in respect to the guidelines in product development, procedures, and programs for ensuring product quality and food safety, product labeling and testing, etc. Based on a recent survey of 462 U.S. northwest regional specialty food processors,[1] about 60% of specialty food processors have less than $500,000 annual sales, and about 71% of them hire less than 50 employees; i.e., the majority of the specialty food processors fall in the small business category. Hence, many of them have limited resources and are seeking information and materials that can help them to be more competitive in the business. In addition, there is a trend of more small entrepreneurs getting into the specialty food business.

This chapter is intended to provide information about where and how to seek additional help and resources, as well as some other key information that may help entrepreneurs who are in the specialty food business, just started, or plan to get into specialty food processing.

Additional Resources in Processing Specialty Food Products

It is clear that success of a specialty food product starts with a great product idea. Unfortunately, over 90% of great product ideas are stopped by the cost of equipment and facilities before the entrepreneurs start since commercial food products must be produced in a licensed production facility. Many entrepreneurs are limited by the equipment, facilities, or even a space to process food on a large enough scale for commercialization. Instead of investing your own processing equipment and facilities, there are some potential alternatives that can help move the product ideas to reality. The following sections discuss the utilization of contract packers, kitchen incubators, and other potential facilities to process specialty foods for commercialization.

Contract Packers or Co-Packers

Contract packers, called co-packers, are an increasingly popular option of having someone else process, package, and possibly market your product. They are the existing food processors to make your product according to your specifications. Co-packers usually sign a nondisclosure agreement, promising not to copy, disclose, or use your recipe or formula except in your own products. You may check Thomas Food Industry Register (http://www.tfir.com)[2] for co-packers in different states.

Depending on your needs, a co-packer can contract with you to produce, pack, store, and distribute your product according to your recipe and specifications. If on-site production facilities or expansion of your current facilities is not possible or feasible, a co-packer may be a choice. Some co-packers let you put your private label on their products and market them as your own. If you are entering a market where your best chance at success is to go in with multiple products, then this might be a good idea. Also, if you are interested in retailing, but have not yet developed your own product, this might be the way to make a name for your company.

Hundreds of companies provide this service. Each has different abilities. The key is to find the right one to meet your specific needs and interest. To determine if a particular co-packer is a good match for you, gather the following information and ask the questions suggested in Table 12.1. Be sure to interview a prospective co-packer for the best match.

- Gather name
- Check for quality
- Check for compatibility
  - Level of involvement
  - Supplies
  - Ingredients
  - Volume
- Case amount

TABLE 12.1
Examples of Questions to Ask When Interviewing a Co-Packer

Subject	Specific Questions to Ask
Volume (batch sizes)	• What is your minimum size co-pack run? • What is the ideal (most cost-efficient) size run?
Products	• What kinds of products can you process? • High acid and acidified? Low acid? • Liquidity limitations? • Container sizes and limitations?
Requirements	• What must my company have in advance before working with you? • A certified commercial-scale processing recipe? Labels? • What is the typical timeframe for setting up a co-pack run?
Ingredients	• Do you source ingredients? • Do you supply ingredients from your inventory? • Can the client provide the ingredients? • Storage? • Do you provide ingredient and glass storage for the client?
Warehousing	• Do you provide warehousing space and services for finished products (such as inventory management, drop shipping, freight arrangements, etc.)? If not, how soon must the client take delivery of finished products?
Marketing and sales	• Do you provide any marketing services for clients (such as joint marketing, broker services, distribution and sales)?
Product development	• Do you have product development capabilities (lab, pilot plant, etc.)? • Can you do nutrient analysis? • Can you do recipe formulation?
Cost	• How do you decide processing charges? Are glass charges extra? • What is a typical run cost? • What do you look for in a client? • A long-term relationship? • A company with growth potential? • A company with compatible products? • Other?

Source. Modified from Mullen, A., G. McWilliam, L. Cooper, J. Delaney-Shirley, E. Ingalls, L. Brushett, and C. Violette, *New Hampshire Specialty Food Producers Handbook and Resource Guide*, New Hampshire Department of Agriculture, Markets and Food and the University of New Hampshire Cooperative Extension and Office of Economic Initiatives, 1996.

Licensed Community Kitchen or Kitchen Incubator

In addition to co-packers, another option is to use a community kitchen or kitchen incubator to start your initial product development. A community kitchen is a shared-use facility where caterers, street cart vendors, farmers, and processors of specialty/gourmet food items can prepare their food products in a fully licensed and certified kitchen. They are often sponsored by an umbrella nonprofit organization or existing business incubator, providing start-up businesses the opportunity to explore food production without the high cost of buying their own equipment or constructing their own building. They also usually offer technical assistance in food production as well as general business management skills, networking opportunities among entrepreneurs, and the opportunity to form shared services cooperatives for marketing, distribution, and supply purchasing.

The licensed community kitchen is a place to provide shared-use kitchen space, and also a place where small business owners can learn more about the technical aspects of food production and receive assistance with skills such as

- Product development
- Ingredient sourcing
- Packaging and labeling
- Shelf-life information
- Nutritional analysis
- Marketing strategies, and more

Kitchen incubators may be used by

- Start-up food businesses in need of their first facility
- Home-based businesses that wish to legalize and grow their operation
- Established businesses that rely on one-off or difficult situation kitchen rentals
- Established businesses that look to grow or reach a new market

These businesses include caterers, personal chefs, bakers, street venders, cake decorators, and producers of specialty food items.

Culinary Incubator (www.CulinaryIncubator.com) is a database website designed to find kitchen incubators in the United States. The website offers direct contact to 116 incubators around the country. In addition, it provides information about kitchen rentals for prospective tenants that can be found on its "Renters Info" page. Current kitchen incubators can sign up and get their kitchen added to the directory free of charge.

Other Facilities

As part of the nationwide Land-Grant University System, the University Cooperative Extension Service provides educational outreach programs to improve people's lives, including education and training programs on a variety of topics in food science and technology. Specific topics may include food processing, product development, food safety, food regulation, and other emerging issues related to food. A list of land-grant universities at each state can be found from the Association of Public and Land-Grant Universities (APLU) (formerly the National Association of State Universities and Land-Grant Colleges, or NASULGC) (https://www.aplu.org/netcommunity/). The information about the specific Cooperative Extension Service program at each of these land-grant institutes can be found through APLU's website. You may contact your local county office for information on the Extension service programs near you. The state's department of agriculture can also direct you to the state Extension programs and the right contact person.

Some of the common Extension services in food science-related programs may include the following:

- Fee-based workshops and short courses in a wide range of topics related to food processing, new technologies, food regulation, food safety, and product development. These can be traditional classroom teaching or online courses.
- Phone conversations or person-to-person meetings with Extension agents or specialists for getting advice on specific topics and questions.
- Education materials provided through Extension fact sheets, newsletters, CDs/DVDs, and more commonly, websites.
- Contracted service. This may include product testing, product development, and specific technologies for solving the problems.

In addition to the Cooperative Extension Service, you may take advantage of other university facilities. Here are two examples:

- Pilot plant facility. Many university food science departments have pilot plant facilities for meeting research and teaching needs. Based on the name, pilot plant facilities are usually small-scale equipment, and are ideal for start-up companies in a small-scale process. Many of these facilities are FDA or USDA inspected and can be utilized for manufacturing commercial products. Entrepreneurs may take advantage of using these facilities during their product development stage through contract. In addition to using the pilot plant equipment and facilities, entrepreneurs may also seek help in product formulation, improvement of processing technologies, and product testing through contract service agreement.

- Food technology or innovation center. Several universities in the United States have food technology or innovation centers with the mission to help small entrepreneurs and start-up companies. Most of these centers offer fee-based contract assistance, from recipes to processing and market. They usually have the pilot plant type of equipment, and function similarly to the licensed community kitchens. Check with your state department of agriculture or county Extension office; they can direct you to the appropriate food technology or innovation centers near you.

Scale-Up from Bench to Commercial Production

Many specialty food products started with a family recipe with a serving size of up to 10 people or so. Scaling up from home recipes and bench type to large-scale commercial production can be challenge. When scaling up, two major issues need to be overcome: modification of recipes and selection of an appropriate production system.

Modification of Home Recipe for Commercial Production

Modification of a recipe is almost absolutely necessary when scaling up for commercial production. Scaled-up batches of a product may be very different from the original formula because of differences in taste, texture, aroma, or appearance. It may be possible to modify the scaled-up formulation so that it more closely resembles the original, but it may not be possible to make it identical. This is not unusual because foods are a complex mix of interacting chemical and physical systems. As these systems become larger, the ways in which they function may change. Therefore, the scale-up process is not necessarily straightforward and may require multiple attempts and trial and error.[4]

Due to the complex interactions of different food ingredients, simplified formulations tend to scale-up more easily and may also be less expensive to produce. Several factors can be used to simplify product formulations, as described in Table 12.2.

Sometimes, it may be necessary and wise to seek external help. Many consultants and lab services offer a range of technical services. Many co-packers also have the facilities and resources available to provide this assistance. Costs can vary widely depending on the firm hired and the complexity of the product. Shopping around and working with a qualified individual or firm is important.

Process and Equipment Scale-up

Major Changes in Process Equipment Requirements[4]   Major changes in process equipment requirements can occur when scaling up from kitchen to production volumes. This is partially due to the physical laws governing the unit processes involved. Heat transfer (cooking and cooling) and mixing in tanks or vessels are two of the most common treatments in

TABLE 12.2
Factors to Consider When Scaling up a Home Recipe for Large Quantity Production

Factors	*Description*
Functionality	Are there other ingredients that do the same job? There are often ingredients with specific functionality that may be used to replace home-style ingredients in scaled-up batches. Bread crumbs can be used as a thickening agent in home recipes. However, in a commercial batch starch may be more appropriate. Starch comes in a variety of forms that can provide the same functionality as the bread crumbs in a more reliable manner over a broader range of conditions.
Form	How similar are the ingredients? A product formula that consists of either a few dry ingredients or a few wet ingredients will probably scale up easier than a formula that contains many different wet and dry ingredients. For example, a formulation for a spice blend might scale up easier than a formula for a gelatin salad. Likewise, formulas that contain ingredients of a similar color might scale up more easily than a product with many different colors. For example, a red beverage might scale up easier than a candy with red, blue, and purple stripes.
Cost	Is a less expensive ingredient available? Sometimes alternative ingredients can be found that are less expensive than those in the original home recipe. For example, weight for weight starch is often a less expensive thickening agent than bread crumbs.
Availability	Will the ingredient be on hand when needed? Many fresh ingredients are seasonal. Items such as herbs are plentiful during warmer months, but may become unavailable when the weather turns colder. Oleoresins and essential oils of herbs are available year-round and offer very similar flavors.
Measurement unit	Converting the recipe to standardized units of weight or volume is necessary to avoid errors. While units such as teaspoons and cups are standardized, they are relatively small and will still need to be converted. Kilograms, pounds, grams, and ounces are common weight measures, and gallons and liters are common volumetric measures. The production process can be streamlined by using these units. However, some items may need to be converted from units of volume to units of weight, depending on how the food will be processed and what equipment is available.

Source. Adapted from Scott, D.D., T.J. Bowser, and W.G. McGlynn, Scaling Up Your Food Process, Fact Sheet FAPC-165, Robert M. Kerr Food and Agricultural Products Center, Oklahoma State University, http://www.fapc.biz/files/factsheets/fapc141.pdf (accessed October 1, 2011).

food processing. Both of these treatments are heavily dependent on the unit surface area (Sa) to volume (Vol) relationship. The ratio of surface area to volume (Sa/Vol) normally changes during the scale-up process. For example, the Sa/Vol ratio is reduced by nearly a factor of 2 when scaling up from a 20-quart kitchen stockpot to a 100-gallon hemispherical tank (hemispherical tanks are common cooking vessels found in commercial kitchens). The decrease in Sa/Vol indicates that there may be a major difference between the mixing and heating characteristics of the two vessels. The ratio becomes much more important for temperature-sensitive and difficult-to-mix materials. Burn-on is often a problem in scale-up, since higher temperatures are needed at cooking surfaces to speed up heat transfer to the product. Table 12.3 lists common scale-up issues for mixing and heating processes.

Batch vs. Continuous Process   When it comes to the specialty food process, there are choices between batch type and continuous process. Each system has its own uniqueness and limitation. Batch processing refers to any food production conducted in discrete batches and is the traditional means of preparing food products. A specified volume or mass of ingredients are combined and undergo processing simultaneously as a unit (such as mixing or cooking) before being transferred to a further processing step (e.g., canning). Continuous processing means that the product is constantly

TABLE 12.3
Common Scale-up Issues for Mixing and Heating

Mixing	*Heating*
Dispersion of ingredients	Burn-on
Mixing time limitation	Heat time limitations
Power requirement	Heat media (steam, electric, hot water, gas, or other)
Materials handling (filling and emptying tank)	Hold time (product quality)
Hold time (product quality)	Cleanup and sanitation
Incorporation of air	Materials handling (pumping, filling and emptying tank)
Cleanup and sanitation	
Changes in physical characteristics of materials (e.g., particle size reduction, color, volume, thickness)	

Source. Adapted from Scott, D.D., T.J. Bowser, and W.G. McGlynn, Scaling Up Your Food Process, Fact Sheet FAPC-165, Robert M. Kerr Food and Agricultural Products Center, Oklahoma State University, http://www.fapc.biz/files/factsheets/fapc141.pdf (accessed October 1, 2011).

moving through the system and is not held "captive" for any length of time. Combinations of batch and continuous processes are common. Both methods have advantages and disadvantages that must be considered on an individual case basis. Normally, continuous processing requires less hold time and handling of a product. Batch processing allows more control of discrete units of production.

Some of the advantages of batch processing may include the following:

- Maximum flexibility
- Easy changeover
- Varying production rates
- Lower capital costs
- Simple operation and control

The disadvantages of batch processing are as follows:

- Higher labor costs
- Higher operating costs and less efficiency for utilities
- Often requires more floor space
- Lower product uniformity

The continuous processing system is usually applied for relatively large-scale processing and feasible only with high production rates. The advantages may be as follows:

- More efficient use of utilities and floor space
- High production rates
- Can produce quick return on investment
- Easy to control processing quality and uniformity
- Lower labor costs

The system has the following disadvantages:

- Lower product flexibility and tolerance to changeover
- High capital cost
- Must run continuously to justify expense

While considering the above specific factors, the decision to choose which process system really depends on the budget, the production rate of the product, and the availability of the system.

Other considerations in the scaling-up process may include product transfer, facility scale-up, waste management, storage, package, rework, inspection, recalls, sanitation program, and supplier certification.[4] Frequently, a contract packager or co-packer will initially be selected to manufacture the product before investing in process equipment until the product is established in the marketplace with a proven sales track record. More detailed information about each of these considerations can be found at Scott et al. [4]

Small-Scale Food Processing Equipment[5,6]

There are many other challenges that the entrepreneurs may have to face due to the nature of the small-scale production and limited resources. The following sections touch some of them, including the consideration for small-scale equipment, quality and resource of ingredients, product package and labeling, and product testing.

Special equipment suitable for small-scale processing often is either unavailable or too expensive for start-up businesses. Entrepreneurs must take the best equipment that is available, and then sequence available units of equipment to work as efficiently as possible. Equipment type, arrangement, and amount are determined by the product being processed, the quantities of each product processed, the size and type of containers into which the food is to be packed, and the need for special operations that impart unique qualities to the food. Food processing equipment must be made from materials that are durable, not facilitate undesirable changes in the color and flavor of foods, not corrode, and be a nontoxin. The preferred material for food contact surfaces is stainless steel.

Food processing equipment is usually classified into the following three categories:[6]

- Low volume: Less than 40 pounds/day (5 gallons), and can be done manually. Use of manual equipment such as specialized cutters may improve speed. Typically hand filled.
- Food service volume: 50 to 150 pounds/day (10 to 25 gallons). They are the restaurant and cafeteria type equipment.
- Food processing volume: More than 200 pounds/day (25 gallons). They are small-scale food processing equipment, and can be difficult to find due to limited demand. This type of equipment is heavy duty, designed for continuous operation. The commercial motors are rated in horsepower, typically 1/4 and higher. Equipment is made of food-grade

materials, in which stainless steel is preferred due to durability and ease of cleaning. It costs hundreds to thousands of dollars per unit.

Some of the common equipment for small-scale production of specialty foods is listed and described in Table 12.4. More information on the small-scale fruit and vegetable processing equipment can be found in Table 5.1 in Fellows (2004).[5]

When purchasing equipment for small-scale operations, the following issues should be considered:

- Can the equipment be justified? Volume must justify equipment need.
- Determine capacity needed in volume per unit of time.
- Understand maintenance and consumables costs before purchasing. In general, the higher the voltage, the cheaper to run.
- Is used equipment an option for reducing the cost? If so, do not overlook the inspection and stay away from scales and electronics without inspection. Consider warranty, condition, availability of spare parts and service, lease option, etc. Also consider renting space in an established kitchen to complement your own equipment.
- Make sure equipment meets FDA/USDA requirements, materials used, and design, and is easy to clean and service.
- Can most mechanical issues be dealt with?
  - Hydraulics, pneumatics, conveying
  - Proper speed for your scale of operation
- Know the equipment specifications and know your facilities' electrical and steam capabilities
- Shop around as much as possible. Get at least three quotes for the same type of equipment, and seek sales references. A deal that sounds too good to be true is probably not.

A searchable directory, Factory Automation Resources (http://www.factory-automationresource.com/category/food-processing/used-food-processing-equipment.html), provides a list of companies in different states selling used food processing equipment.

Food and Other Functional Ingredients

Ingredients directly impact the final product quality, safety, and price. For specialty foods, some of the unique characteristics about food ingredients are natural, organic, locally produced, and sustainable. For ensuring final

TABLE 12.4
Common Food Processing Equipment for Small-Scale Production

Preparation Equipment

Size reduction	Cutters, grinders, slicers. Basically large food processors. Continuous feed is preferred, stainless steel or plastic food contact parts, FDA or USDA approved for food, at least 1/2 HP. Examples: Vertical cutter-mixer used to chop, mix, blend, puree, or emulsify, operates normally in batch mode; shaft type mixer-cutter used to mix, puree a fixed volume, portable.
Mixer	May include dough mixer, food processor, blender, pumps. Dough mixers from 12 to 140 quarts.
Pulper/finisher	Puree fruit/vegetable, separate seeds and skin.
Juice extractors (from puree)	Squeezing action. Models include hydraulic plate and frame, accordion, belt press, and screw press.
Peeler	Abrasive type is good for round homogeneous products, such as potato peeler.

Heating Equipment

Kettles	One of the most common pieces of equipment in small-scale processing. Available from 0.5 to 200+ gallons. Steam is supplied or generated in place (self-contained) by electricity or gas. Preferable options include 316 stainless steel, tilting, bottom discharge with valve, and agitator. Important to cook a full load to avoid scorching. Steam kettles can be used for blanching, cooking, and pasteurizing.
Continuous pasteurizer	Used for flow type of products. Types include tube in shell, plate, direct steam injection, UV treatment.
Oven/dryer	For baking, roasting, and drying. Forced-convection oven is faster. Combination oven-steamers are available.
Fryer	10- to 40-gallon capacity is typical. There are gas-powered or electric-powered fryers. Gas fryers may be cheaper to operate and recover heat more quickly. Also, high-end gas fryers heat up faster than any other type. Gas fryers can use natural or propane gas. Electric fryers are very efficient and feature more safety measures and enhanced controls for temperature and time limits. With electric fryers, a low-watt design may be available that extends the life of both the heating elements and the oil or shortening. They also disassemble quickly for easy portability and storage.

Cooling Equipment

Refrigerator	Good temperature control. Built-in temperature indicator preferred.
Freezer	Blast freezer for quick chilling gives best quality, but is more expensive than conventional units.

Filling/Packaging Equipment

Dry products	Usually filled by weight manually or with auger type fillers.
Wet products	Filled by volume or weight with a variety of fillers based on product consistency. If the product is pumpable, filling can be done by gravity or using piston fillers. Capacity is given by the flow rate or containers per minute. Some units can be upgraded by adding extra filling heads.

(Continued)

TABLE 12.4 (CONTINUED) Common Food Processing Equipment for Small-Scale Production	
Bag sealers	Different types are available, including manual, form-fill-seal, gas flushed, and vacuum sealers.
Tray/cup sealers	Lid or film is heat sealed to container. Can be done manually or fully automated.
Steam or vacuum cappers	Normally used for glass jars to decrease oxygen in headspace and to provide vacuum.

Source. Modified from Northeast Center for Food Entrepreneurship at the New York State Food Venture Center, Cornell University, Small Scale Processing Equipment: Fact Sheets for the Small Scale Food Entrepreneur, 2007, http://necfe.foodscience.cornell.edu/publications/pdf/FS_SmallScaleProcessingEquip.pdf (accessed October 1, 2011).

product quality and food safety (microbial and allergic safety), some of the basic considerations are described below:

- Each ingredient should have a specification sheet from the supplier.
- A specification sheet should give details about the characteristics of the ingredients and quality attributes.
- It may be necessary sometimes to work with a supplier to establish a specification for an ingredient that meets your needs.
- It is your responsibility to monitor your incoming ingredients to ensure that they meet the specification standards that you expect.
- It is not cost-effective to test every ingredient that is used in your product each time you receive it.
- You need to identify the critical or sensitive ingredients that will be monitored 100% or most frequently.
- The 80/20 rule is useful to follow. It is a probability distribution law. Twenty percent of ingredients will cause 80% of your ingredient concerns. The remaining ingredients should have a limited auditing schedule to ensure quality.
- The ingredients and quality of those ingredients will help define your product quality.
- If an ingredient does not meet your quality standard it should be returned to the supplier.

In addition to the major food ingredients, there are many functional ingredients that are used for enhancing quality, extending shelf-life, and ensuring microbial safety of food products. Table 12.5 lists the major functional ingredients and their functionality.

TABLE 12.5
Common Functional Food Ingredients and Their Functionalities

Types of Ingredients	Functions	Examples of Uses	Names Found on Product Labels
Chemical preservatives	Prevent food spoilage from bacteria, molds, fungi, or yeast; slow or prevent changes in color, flavor, or texture and delay rancidity; maintain freshness	Fruit sauces and jellies, beverages, baked goods, cured meats, oils and margarines, cereals, dressings, snack foods, fruits and vegetables	Ascorbic acid, citric acid, sodium benzoate, calcium propionate, sodium erythorbate, sodium nitrite, calcium sorbate, potassium sorbate, BHA, BHT, EDTA, vitamin E
Sweeteners	Add sweetness with or without the extra calories	Beverages, baked goods, confections, table-top sugar, substitutes, many processed foods	Sucrose, glucose, fructose, sorbitol, mannitol, corn syrup, high-fructose corn syrup, saccharin, aspartame, sucralose
Color additives	Offset color loss due to exposure to light, air, temperature extremes, moisture and storage conditions; correct natural variations in color; enhance colors that occur naturally; provide color to colorless and "fun" foods	Many processed foods, such as candies, snack foods, margarine, cheese, soft drinks, jams/jellies, gelatins, pudding, and pie fillings	FD&C Blue Nos. 1 and 2, FD&C Green No. 3, FD&C Red Nos. 3 and 40, FD&C Yellow Nos. 5 and 6, Orange B, Citrus Red No. 2, annatto extract, beta-carotene, grape skin extract, cochineal extract or carmine, paprika oleoresin, caramel color, fruit and vegetable juices
Flavors and spices	Add specific flavors (natural and synthetic)	Pudding and pie fillings; gelatin dessert mixes, cake mixes, salad dressings, candies, soft drinks, ice cream, BBQ sauce	Natural flavoring, artificial flavor, and spices
Flavor enhancers	Enhance flavors already present in foods (without providing their own separate flavor)	Many processed foods	Monosodium glutamate (MSG), hydrolyzed soy protein, autolyzed yeast extract, disodium guanylate or inosinate
Fat replacers (and components of formulations used to replace fats)	Provide expected texture and a creamy mouthfeel in reduced-fat foods	Baked goods, dressings, frozen desserts, confections, cake and dessert mixes, dairy products	Olestra, cellulose gel, carrageenan, polydextrose, modified food starch, micro-particulated egg white protein, guar gum, xanthan gum, whey protein concentrate

(Continued)

TABLE 12.5 (CONTINUED)
Common Functional Food Ingredients and Their Functionalities

Types of Ingredients	Functions	Examples of Uses	Names Found on Product Labels
Nutrients	Replace vitamins and minerals lost in processing (enrichment), add nutrients that may be lacking in the diet (fortification)	Flour, breads, cereals, rice, macaroni, margarine, salt, milk, fruit beverages, energy bars, instant breakfast drinks	Vitamin B_2, niacin, niacinamide, folate or folic acid, beta-carotene, potassium iodide, iron or ferrous sulfate, alpha-tocopherols, ascorbic acid, vitamin D, amino acids
Emulsifiers	Allow smooth mixing of ingredients, prevent separation; keep emulsified products stable, reduce stickiness, control crystallization, keep ingredients dispersed, and help products dissolve more easily	Salad dressings, peanut butter, chocolate, margarine, frozen desserts	Soy lecithin, mono- and diglycerides, egg yolks, polysorbates, sorbitan monostearate
Stabilizers and thickeners, binders, texturizers	Produce uniform texture, improve mouthfeel	Frozen desserts, dairy products, cakes, pudding and gelatin mixes, dressings, jams and jellies, sauces	Gelatin, pectin, guar gum, carrageenan, xanthan gum, whey
pH control agents and acidulants	Control acidity and alkalinity, prevent spoilage	Beverages, frozen desserts, chocolate, low-acid canned foods, baking powder	Lactic acid, citric acid, acetic acid, ammonium hydroxide, sodium carbonate
Leavening agents	Promote rising of baked goods	Breads and other baked goods	Baking soda, monocalcium phosphate, calcium carbonate

Anticaking agents	Keep powdered foods free flowing, prevent moisture absorption	Salt, baking powder, confectioner's sugar	Calcium silicate, iron ammonium citrate, silicon dioxide
Humectants	Retain moisture	Shredded coconut, marshmallows, soft candies, confections	Glycerin, sorbitol
Yeast nutrients	Promote growth of yeast	Breads and other baked goods	Calcium sulfate, ammonium phosphate
Dough strengtheners and conditioners	Produce more stable dough	Breads and other baked goods	Ammonium sulfate, azodicarbonamide, L-cysteine
Firming agents	Maintain crispness and firmness	Processed fruits and vegetables	Calcium chloride, calcium lactate
Enzyme preparations	Modify proteins, polysaccharides and fats	Cheese, dairy products, meat	Enzymes, lactase, papain, rennet, chymosin
Gases	Serve as propellant, aerate, or create carbonation	Oil cooking spray, whipped cream, carbonated beverages	Carbon dioxide, nitrous oxide

TABLE 12.6
Example of Ingredient Log Sheet

Ingredient Name	Date Received	Case Code	Quantity	Temp	Brix	pH	Color	Flavor	Appearance	Receiver

Remember that record keeping of all ingredients is essential for quality control of final products. Table 12.6 is an example of an ingredient test log sheet. These log sheets define the acceptable test ranges.

Several resources may be used for finding common functional ingredients. Here are several examples:

- Institute of Food Technologists (IFT) ingredient guideline, which is searchable from http://buyersguide.ift.org/cms/?pid=3001&categoryId=1
- Foodingredients.com at http://www.food-ingredients.com
- Food Ingredients Online at http://www.foodingredientsonline.com

Packaging and Labeling

Packaging Considerations[7,8]

Whether introducing a new product or refreshing an existing line, packaging plays a key role in the success of sales. Food package is as important as the product inside the container. The functions of a package include the following:

- To contain foods and hold the contents and keep them clean and secure without leakage or breakage until they are used
- To protect foods against a range of hazards during distribution and storage, such as to provide a barrier to dirt, microorganisms, and other contaminants, and to protect against damage caused by insects, birds, and rodents, heat, oxidation, and moisture pickup or loss
- To give convenient handling throughout the production, storage, and distribution system, including easy opening, dispensing, and resealing, and being suitable for easy disposal, recycling, or reuse
- To enable the consumer to identify the food, and give instructions so that the food is stored and used correctly

The packaging must not only communicate company philosophy and brand identity to the consumer, but also protect its contents and merchandise well.[7] Several factors should be considered when designing a food package and in selecting the appropriate container. Here are some key questions to ask:[8]

- Does the package describe and enhance the product, establish the brand?
- Does the package allow consumer to see the product and appeal to your target consumers?

- Does it set your product apart from the competition and establish a niche in the marketplace?
- Does the package provide a contamination barrier?
- Is the shape of the package distracting?
- How easy can a label be applied on the package?
- Is the package easy to use?
- How will the package stack for shelf display?
- Does the package conform to federal and state laws?
- Is the cost of the package prohibitive?
- Is the package material recyclable?

There are a wide range of packaging materials and container types to select from, such as glass, plastic, cellophane, paper, cardboard, wood, and metal cans. The selection of a container type depends on many factors, but at a minimum should protect the product from contamination and should enhance its best selling features. Odd-sized containers should be avoided. As a general rule the container should fit and stack on standard store shelves. The selling price that best fits the market will influence the size of the container selected. For example, a smaller container (8-ounce jar) will likely sell faster than a larger size (16-ounce jar), resulting in faster repeat sales. Note that many retailers insist the food products carried in their stores are tamper resistant. These can be simple ribbons, seals, or bows that can be integrated attractively into the package and label design. When broken, these items indicate the package has been opened.

The food business is seeing a packaging revival through changing graphics, distinctive marketing strategies, and more environmentally friendly packaging. Some recent trends of packaging for specialty foods are as follows:[7]

- Efficient packaging with simplified, straightforward graphics and minimal design. One benefit of this can be to project the image of a cleaner, more back-to-basics product inside.
- Planet-friendly packaging using sustainable/green packaging materials, such as less plastic to produce water bottles, natural inks to print label graphics, or recycled paper for labels.
- Functional packaging using containers that have multiple uses, especially products that simplify shoppers' hectic lives.

Package Labeling

All food items packaged must be properly labeled prior to sale according to Code of Federal Regulations (CFR) Title 21, Parts 100–169. *Food Labeling Guide* can be found at the FDA's website: http://www.cfsan.fda.gov/label.html. It is the food processor's responsibility to keep current with the legal requirements for food labeling. The law requires the following information to be on all food items:

- Statement of identity (the common or usual name of the food product).
- The name, street address, city, state, and zip code of the manufacturer, packer, or distributor.
- An accurate statement of the net amount of food in the package.
- The ingredients in the food. If a product is made from two or more ingredients, each ingredient must be listed in descending order of predominance.

Additional information may be optionally included on the package labeling:

- The brand. It is optional to include a brand logo, but for helping market a product and establish a brand in the marketplace, it is important to define and include a brand logo. That will quickly help consumers to find the products and come back to buy them.
- Product dating (manufacturing date, expire date, best by date).
- Instructions for preparing the product.
- Storage instructions or instructions on storage after opening.
- Examples of recipes in which the product can be used.
- A bar code.

The Nutrition Labeling and Education Act amended the Food, Drug, and Cosmetic Act to require most foods to have nutrition labeling. Although the final regulations have been published, they are often changed or updated. The FDA's publication *A Food Labeling Guide* (http://www.cfsan.fda.gov/~dms/flg-toc.html) lists the requirements of the Nutrition Labeling and Education Act; the Food, Drug, and Cosmetic Act; and the Fair Packaging and Labeling Act. It includes a summary of the required statements that must appear on the labels of foods produced in the United States. After the labels are designed and written, it is highly recommended to submit them for review and comments to the regulatory services. This step is not required by law; however, it can save you time and money.

Table 12.7 summarizes the FDA's food package labeling requirements. A food product is considered misbranded if its label makes unwarranted claims for the product or fails to make all the statements required. Misbranding can result in civil and criminal penalties, seizure, and injunction.

TABLE 12.7
Summary of the Information Required to Appear on a Food Package Label Based on FDA Regulation

Required Information	Description
Principal display panel of the food container	Certain required statements must be placed on what is called the principal display panel (PDP). The PDP portion of the label is the area most likely to be seen by the consumer at the time of purchase. For food products standing on a shelf, the PDP is typically the front panel. For those products stacked in a refrigerated case, the PDP is usually the top panel. Statements required on the principal display panel include product identity or name of the food and net quantity or amount of the product.
Statement of product identity	Product identity is the truthful or common name of the product. An identity statement consists of the name of the food and should appear in prominent print or type. The type size should be at least one-half the size of the largest print appearing on the PDP. Common or usual names such as "raspberry jam" or "bean soup" should be used. A descriptive or fanciful name is permitted if the nature of the food is obvious, but should not be misleading. If the food is subject to a standard of identity it must bear the name specified in the standard of identity. A description of the form of the food must be used if the food is sold in different forms, such as sliced or unsliced, whole, halves, etc.
Statement of net quantity	Net quantity or amount should be distinctly displayed on the bottom one-third of the label on the PDP, in one line or several lines parallel with the base of the container. Select a print style that is prominent and easy to read. Only the weight of the food, not the container and wrapping, should be calculated in the net quantity. Net weight should include all ingredients, including water or syrup used in packing the food. Net weights must be stated in both units of the U.S. customary system (inch/pound) and metric measure (meter/gram). Dual declaration in both ounces and the largest whole unit (pounds/ounces or quarts/ounces) is optional. Note that dry and liquid products are measured differently.
Information panel	It is not necessary for all required information to be on the "front" of the principal display panel. The ingredient list along with name and place of business may appear on the information panel. This panel is the side to the right of the PDP. If, due to the package shape, there is no room immediately to the right of the PDP, for example, it's flat like a chocolate bar, then the information panel may be on the back of the package.
Ingredient list	Ingredients must be listed in descending order of predominance by weight in type size at least 1/16 inch in height. Always list the common or usual name rather than the scientific name. Added water is considered to be an ingredient and must be identified. Approved chemical preservatives must be listed, using both the common name and a statement specifying the ingredient is a preservative. The exact function of the preservative may also be included. Incidental additives that have no function or technical effect in the finished product need not be declared. Approved artificial food colors must also be stated by name. The only ingredient where the words *and/or* can be used is oils.

TABLE 12.7 (CONTINUED)
Summary of the Information Required to Appear on a
Food Package Label Based on FDA Regulation

Required Information	Description
	All components of ingredients must be specifically listed with the exception of spices and natural flavors, which can be declared by their common name or simply by "spices" or "natural flavors."
Name and place of business	The name and place of business of the manufacturer, packer, or distributor must appear next to the ingredients statement in type size at least 1/16 inch high. The street address must appear unless it can be readily found in some public document, such as a telephone book or city directory.
Nutrition Labeling and Education Act	Nutritional information is required on all food products with the following exceptions: • Restaurant and deli foods • Infant formula • Medical foods • Bulk foods intended for repackaging • Foods that contain insignificant amounts of nutrients, such as spice blends and coffee • Low-volume products bearing no nutritional claims Low-volume food products processed by small businesses may be exempted from nutritional labeling requirements if they meet all of the following criteria: • For each product, the company must apply to the FDA annually to obtain a small business exemption if the above criteria are met. If the company employs less than 10 full-time employees and the product has sales of less than 10,000 units per year, it does not have to file for the exemption. If the exemption does apply, be sure to contact the FDA to obtain a Small Business Food Labeling Exemption form at http://www.fda.gov/ora/inspect_ref. • If nutritional information is provided, it must follow a defined format and include specified nutrients. The nutritional panel may vary according to the size of the package. Foods sold in very small packaging (less than 12 square inches of total available labeling space) may omit the nutritional label but must include a statement and address where nutrition information can be obtained. 1. The product provides no nutrition information and makes no defined health claims. 2. The firm claiming the exemption has less than the equivalent of 100 full-time employees. 3. During the previous 12 months, less than 100,000 units were sold or it is anticipated that less than 100,000 units will be sold during the period for which an exemption is claimed. 4. The exemption must be claimed prior to the period for which it is to apply.

(Continued)

TABLE 12.7 (CONTINUED) **Summary of the Information Required to Appear on a Food Package Label Based on FDA Regulation**	

Required Information	Description
	5. If, after filing an exemption, either the number of employees or volume of product sold annually increase and your exempt status is lost, you have 18 months to bring your label into compliance with the nutrition labeling requirements.
Nutritional claims	If a nutrient content claim is made, such as "low fat" or "reduced calories," a nutritional panel is required to support that claim, regardless of product sales volume. Terms used on the label, such as *healthy, free, less, high, more, low, good source, light, reduced, fewer, lean,* and *extra lean,* must meet FDA definitions.
	No statements or symbols are allowed to imply unauthorized nutrient claims. This includes heart vignettes that may imply "healthy" unless the vignette is clearly used in another context. Nutrition labeling on retail bulk foods is also required. Prominently displayed nutritional information on raw fruits, vegetables, and seafood at the point of sale is voluntary.

Source. Adapted from FDA, A Food Labeling Guide, http://www.cfsan.fda.gov/~dms/flg-toc.html.

Food Allergy Labeling

Food allergies affect about 2% adults and 4–6% children in the United States, and the number of young people with food allergies has increased over the last decade, according to a recent report by the Centers for Disease Control and Prevention (CDC; http://www.cdc.gov/). Children with food allergies are more likely to have asthma, eczema, and other types of allergies. The FDA issued the Food Allergen Labeling and Consumer Protection Act of 2004 (Public Law 108-282, Title II). Eight major food allergens are described in Table 12.8. The act requires that major allergenic ingredients in food are accurately labeled on packaging.

Allergens must be declared when present as either

- An ingredient
- An ingredient of a compound ingredient
- A food additive or component of a food additive
- A processing aid or component of a processing aid

The package label should carry an allergy warning statement, such as "May contain ..." or "Manufactured in a facility ...." Allergens may appear on a label in one of three ways:

TABLE 12.8 The "Big 8" Food Allergens	
Milk	Includes ice cream, powdered milk, evaporated milk, yogurt, butter, cheese, cream and sour cream, nondairy products, and any other food products containing lactose, caseinate, potassium caseinate, casein, lactalbumin, lactoglobulin, curds, whey, milk solids.
Egg	Egg is present in most processed food and is present if the label indicates any of the following additions: constituent egg proteins or their derivatives (e.g., albumen, ovalbumen, globulin, ovomucoid, vitelin, ovovitelin, silicoalbuminate).
Fish	Such as bass, flounder, and cod. Any type of fin fish; any food product containing fish.
Crustacean shellfish	Such as crab, lobster, or shrimp. Also includes any food product containing these crustaceans.
Tree nuts	Such as almonds, pecans, or walnuts. Also includes any food containing tree nuts (e.g., salads, entrees, cookies, cakes, candies, pastries, or breads).
Wheat	All types of wheat flour; any baked products and any prepared products containing wheat flour, wheat gluten, or wheat starch.
Peanut	Also includes any food product containing peanuts.
Soybean	Also includes any food product containing soybeans.

1. List on principal display panel or in ingredient list, e.g., crabmeat
2. List in ingredient list next to ingredient that does not disclose what it is, e.g., flour (wheat), casein (milk)
3. List after the word *contains*, e.g., Contains: egg, soy, wheat

Product Testing

Part of product development will include some analytical tests. These tests help determine product quality, safety and shelf-life. Tests needed vary with the type and nature of the food product. Check with state and federal regulatory agencies to decide which tests are required for your products. The common tests may include the following:

- Basic quality: pH, sugar, moisture content, water activity, color, texture
- Microbial quality: Total plate count, coliform, yeast, mold, specific pathogens
- Sensory quality: Consumer likeness and preference, specific sensory attribute by using trained panelists

Table 12.9 describes the common tests for quality control of final food products. Some of these tests may be done by the processors in the plant with

TABLE 12.9
Common Tests for Quality Control of Food

Name of Test	Description
pH	Determine the pH or acidity level of a food product. Foods with a low pH contain enough acidity to block the growth of harmful bacteria, therefore reducing the possible accumulation of toxins, especially the deadly *Clostridium botulinum*. Special regulations apply to low-acid canned (pH > 4.6, water activity (a_w) > 0.85) and acidified foods.
Titratable acidity	Approximation for total acidity. It measures the organic acid content of a sample. The procedure is based on titrating the original acid sample with a standard NaOH (sodium hydroxide) solution to the endpoint of pH = 8.2, as well as a magnetic stirrer to allow more consistent and thorough stirring during the titration.
Moisture content	Amount of water in a product. Moisture content measurement involves separating water from a food sample and measuring the weight loss, and is calculated as a percentage of the original (wet) or final (dry) sample weight.
Water activity	A measure of free water in a product available for microbial growth with a range from 0 to 1. Growth of microorganisms in foods can be controlled by reducing the water activity, i.e., drying or binding the water in the food to make it unavailable to microorganisms, e.g., adding salt or sugar. Water activity can be measured using a water activity meter.
Brix (total soluble solid content)	% Soluble solids = weight sugar/(weight sugar + weight H_2O) ° Brix = dissolved sucrose to water in a liquid = g sucrose/(g sugar + g H_2O) Total soluble solids content can be measured by a refractometer.
Microbial analysis	Primarily used for food safety regarding food-borne infections, food spoilage, and food poisoning.
Total plate count	Measure bacterial populations. Bacteria require nutrients, moisture, and favorable temperature to grow. Bacteria may grow in the presence or absence of oxygen. Foods dried or preserved with salts or sugars have minimal moisture. This lack of moisture helps prevent bacteria growth. Acid also prevents the growth of bacteria. Foods with a low pH usually don't require bacteria analysis.
Mold count	Measure mold population. Molds grow well at room temperature, in long-term storage of foods such as peanuts and grains, and in short-term storage of breads without preservatives. A serious problem can develop when molds produce a mycotoxin. Ideal mold conditions: • Temperature of 77–86°F, damp, dark conditions • Acid, neutral, or sweet foods (pH 5.5 or less)
Pathogen tests	*Listeria monocytogenes*: Vegetables, milk, cheeses, fermented meat. *Yersinia entercolitica*: Raw milk, chocolate milk, water, pork, and other raw meats. *Campylobacter jejuni*: Raw milk, cake icing, eggs, poultry, rare beef, water. *Vibrio cholerae*: Water, crab, shrimp, oysters. *Escherichia coli*: Ground beef, raw milk, chicken.

TABLE 12.9 (CONTINUED)
Common Tests for Quality Control of Food

Name of Test	Description
Pesticide and residue analysis	Recommended to test for foods exposed to pesticides or other toxic substances. Some foods contain naturally occurring toxic substances.
Food additives analysis	Determine the level of chemical preservatives added to foods to retard spoilage.
	If a product involves the purchase of prepackaged ingredients, you may want to consider this test.
Shelf-life	Predict food deterioration during normal storage conditions. It helps determine protective packaging requirements for a given food. The prediction of storage life can help eliminate overpackaging, which can be wasteful both financially and environmentally.

a small investment in the instruments, such as a pH meter for measuring pH, refractometer for measuring Brix, and water activity meter for measuring water activity. It is especially important to own a pH meter if you make acidified foods, and a refractometer if making jellied products. However, some other tests, especially the pathogens, chemical residues, etc., are much more difficult to do, and require some skills and a lab setting; thus using outside resources may be a better choice. These outside testing resources may include the following:

- Certified test labs
- State agencies service labs
- University labs that provide testing services

Your state agencies or Extension county offices may be able to direct you to the right facilities for your sample testing.

Summary

The specialty food business is a very complex system, and involves many different aspects, from specific product concept, processing technologies, packaging, and specific controls for ensuring product quality, food safety, and shelf-life, to licensing, marketing, etc. It requires a good understanding and broad knowledge in food manufacturing, business development and management, and legal issues. The good news is that many resources are available to assist the success of your business, from the local and state to federal levels. There are also many education programs available offered by the state department of agriculture or commerce, Land-Grant University Extension Service, nonprofit organizations, and commodity groups. The key is to take time doing enough homework before starting.

References

1. Zhao, Y., and Daeschel, M. 2009. A report of US northwest specialty food processors' survey.
2. Thomas Food Industry Register. http://www.tfir.com.
3. Mullen, A., McWilliam, G., Cooper, L., Delaney-Shirley, J., Ingalls, E., Brushett, L., and Violette, C. 1996. *New Hampshire specialty food producers handbook and resource guide*. New Hampshire Department of Agriculture, Markets and Food and the University of New Hampshire Cooperative Extension and Office of Economic Initiatives.
4. Scott, D.D., Bowser, T.J., and McGlynn, W.G. Scaling up your food process. Fact Sheet FAPC-165. Robert M. Kerr Food and Agricultural Products Center, Oklahoma State University. http://www.fapc.biz/files/factsheets/fapc141.pdf (accessed October 1, 2011).
5. Fellows, P. 2004. Small-scale fruit and vegetable processing and products: *Production methods, equipment and quality assurance practices*. Vienna: United Nations Industrial Development Organization.
6. Northeast Center for Food Entrepreneurship at the New York State Food Venture Center, Cornell University. 2007. Small scale processing equipment: Fact sheets for the small scale food entrepreneur. http://necfe.foodscience.cornell.edu/publications/pdf/FS_SmallScaleProcessingEquip.pdf (accessed October 1, 2011).
7. Denis, N.P. Food garb. *Specialty Food Magazine*. http://www.specialtyfood.com/news-trends/featured-articles/retail-operations/food-garb (accessed October 1, 2011).
8. Idaho State Department of Agriculture, Marketing Idaho Food and Agriculture. 2008. Starting a specialty foods business, 2008. VIII ed. http://www.agri.idaho.gov/Categories/Marketing/Documents/specialtyfoodbook.pdf (accessed October 1, 2011).

Index

Ќ